Nature,
Mother of Invention

Felix R. Paturi

Nature, Mother of Invention

The Engineering of Plant Life

Translated from the German
by Margaret Clarke

Harper & Row, Publishers

New York, Hagerstown, San Francisco, London

Translated from the German, *Geniale Ingenieure der Natur*,
by Margaret Clarke.
© Econ Verlag GmbH, Düsseldorf und Wien, 1974.

NATURE, MOTHER OF INVENTION.

FIRST U.S. EDITION

ISBN: 0-06-013288-4

LIBRARY OF CONGRESS CATALOG CARD NUMBER: 74-1846

PRINTED IN GREAT BRITAIN

Contents

The marginal numbers beginning on page 15 refer the reader to the illustrations. The following refer to figures in the text: 29 (p.48), 37 (p.55), 38 (p.57), 47 (p.63), 82 (p.158), 85 (p.185).

ACKNOWLEDGMENTS

Acknowledgments are due to the following for permission to reproduce illustrations:

Botanical Garden and Museum, Berlin-Dahlem (18); British Information Services, London (8); Desert Botanical Garden, Phoenix, Arizona (13); Deutscher Aeroclub, Frankfurt (41, 42); Dow Chemical GmbH, Frankfurt (69); German Federal Ministry of Defence (87); Dr Y. Haneda (100a and b); Hexcel Honeycomb, Dublin, California (24); IBM, Stuttgart (79, 80, 81); Mrs Aenny Kessler (11, 73); National Gallery, London (32); Neumann Verlag, Radebeul (85, 90); Photopress Grenoble (59); Albert Pietsch (22); Professor W. Rauh (74); Professor Hans Reinerth (17); Sames Electrostatic GmbH, Darmstadt (59); Springer Verlag, Berlin (37, 38); Städtische Sukkulentensammlung, Zurich (14); Verlag Gebrüder Bornträger, Stuttgart (26, 28). All other illustrations are from the author's collection.

CHAPTER ONE

Man the Extravagant

A single farmer in the USA today provides an average of forty people with foodstuffs and textile fibres. A Chinese rice farmer, on the other hand, works strenuously from dawn to dark without securing more than his own bare existence. What a difference in efficiency! In efficiency? Yes, but it is precisely in relation to efficiency that the activity of the American farmer, critically considered, comes off worse than that of the Chinese peasant. The US farmer's harvest is many times greater than that of his Asiatic colleague, but what price does he pay for this?

Efficiency is the relation between output and expenditure. If we convert the energy gained from the harvested plants into kilowatt hours and compare it with the energy expended for that harvest, the result is startling: for 50 harvested energy units the American farmer invests 250 fuel energy units, the Chinese farmer only a single unit of human energy. This means simply that the primitive countryman of the East works at an efficiency rate of 5000 per cent, and the US farmer, equipped with the most advanced technical aids, at an efficiency rate of only 20 per cent.

Certainly they represent the two extremes. The American ruthlessly exploits energy, the Chinese ruthlessly exploits himself. It is true that the ruthless exploitation of energy guarantees a more agreeable way of life; but for how long? In a few generations we shall have completely exhausted our available reserves of mineral energy. The oil reserves will have dried up; the coal seams will have been worked out. Man will have to cross off his list of natural resources such energy suppliers as petrol, diesel and fuel oil, plus the aircraft propellant kerosene, as well as the material products from coal and petroleum: tar and asphalt for street-paving; synthetic fibres, like polyester, nylon and acetate, synthetic rubber for the tyre industry – in short, all the raw materials of the plastics industry.

The ruthless exploitation of energy and materials that we permit ourselves gallops faster today than any inflation. When

7

a housewife weighing 130 lb goes shopping in a car weighing more than a ton and consuming high-grade petrol at a rate of 20–25 m.p.g., in order to bring home two or three paper bags of provisions, the operation can scarcely be termed a prudent husbanding of our limited energy resources. Every American uses up quantities of energy corresponding, on the average, to the bodily labour of 500 human slaves. Australians make use of 250 such energy slaves per head, Europeans of over 200, South Africans of 125 or so.

If, as citizens of a Western industrial state, we believe that we cannot live without the energy which is today at our command, simply considering the world distribution of energy may serve to dissipate these rather one-sided conceptions. For every American at the present moment, energy is being consumed every year, directly or indirectly, equivalent to that of 12·7 tons of coal. The Canadian in comparison uses 11·0 tons. Swedes are in the third place with 7·2 tons. Then follow the East Germans (6·9 tons), Belgians and Danes (6·7 tons); the British (6·5 tons) and the citizens of West Germany and the Netherlands (6.2 tons). An Ethiopian nomad, on the other hand, could carry his annual energy equivalent on his back in a 70 lb sack.

Now it might be supposed that the near-limits of energy consumption per head had been reached in the Western world. But they have not. American scientists allow for a population increase of 30 per cent in their country between 1970 and 2000. The increase in energy consumption, however, for the same period, is estimated at 400 to 500 per cent!

Let us revert for a moment to the image of the 'energy-slave'. Dr James P. Lodge Jr, of the National Atomic Research Centre in Boulder, Colorado, takes the term literally when he says: 'Of course we must limit our population growth, but it would be still more helpful to cut down on the numbers of our energy-slaves. That this demands a high degree of rethinking and a new set of priorities is obvious.'

Certainly the rethinking will be difficult, but let us hope it will not prove impossible. If we do not manage it in time, and that means within the next five to ten years, these energy-slaves with their waste products – far more difficult to dispose of than human excrement – will get entirely out of control and become more and more toxic, till one fine day they abandon us as butlers desert impoverished noble families.

Scarcely more sparing than the use of energy has been that of materials. We reduce our natural raw materials in the manufacture of refrigerators, beer bottles, newspapers, and an indescribable variety of *de luxe* packaging, use them for a brief space, and then throw them on the rubbish tip or burn them up in high-temperature plants specially designed for the purpose. The result of this prodigal use of materials is not difficult to foresee.

Towards the end of our millennium there will no longer be any natural paper. Synthetic paper will take its place, but the basic material of synthetic paper, petroleum, will likewise be exhausted by the coming generations. Even before the year 2000, reserves in silver, aluminium, mercury, lead, platinum, zinc and tin will be used up. By the beginning of the third millennium, when perhaps our grandchildren and great-grandchildren will be living, there will be no chromium, no nickel, no tungsten and no iron.

Scarcity of energy and dearth of materials are two spectres threatening the near future, but they are neither the only nor the worst consequences of unlimited prodigality. Dearth of material on the one hand causes superfluity on the other: superfluity of filth. Rubbish tips devour the landscape, poison the ground water, become breeding-grounds for new types of pestilence. The UN World Health Organization has issued a warning that humanity is threatened by a growing danger, that of veritable explosions of plague, especially in view of the breakneck growth of towns. In almost every country in the world there are more rats than men, even in the USA. Garbage is their staple diet, and as long as we do not master the garbage problem we shall not get rid of rats, nor be able to eliminate the possibility of new plagues.

Still more dangerous than garbage, which can be tipped onto a waste-heap, is the ever-present 'invisible garbage': dust, exhaust fumes and sewage. These are not just a possible danger in the future, but already a scourge here and now. They are transforming rivers, lakes and seas into cesspools, biologically dead; they are contaminating field and pasture, and poisoning the atmosphere. In the centre of Tokyo there are already kerbside pumps dispensing oxygen. For a coin in the slot, the modern inhabitant of this metropolis receives a lungful of oxygen, when dizzied by sulphur dioxide, lead and carbon monoxide, and on the verge of a blackout. Will such pumps be a commonplace in the world of our children?

The latest statistics are horrifying: the amount per head of exhaust fumes and dust poured into the air of West Germany in the course of a year weighs more than the domestic refuse per head which accumulates or is scrapped during the same period. The figures are: eight million tons of highly poisonous carbon monoxide, four million tons of no less noxious sulphur dioxide, four million tons of dust and rust to poison the lungs and larynx, two million tons of nitric oxides, and as many tons of hydro-carbons.

Pollution by gas and dust increases at the same rate as the consumption of energy, for both are the products of combustion. And the consumption of energy grows and grows – for how long? If it bears out the projections of the scientists, and the inevitable world energy crisis comes upon us, it will lead straight to the most terrible of all conclusions: world war. Not without a struggle will the nations allow the last reserves to be peacefully distributed.

For some 40,000 years *Homo sapiens*, knowing and thinking man, has been on the earth. For some 5,000 years he has lived in advanced civilizations and known cities. The development of industry does not go back further than 200 years; only since then has disaster been overtaking us.

Life is not, in the last resort, mere metabolism and energy exchange, so it seems natural to consume energy and use up material proportionately to the 'intensity' with which we live. The abiding question is, how has thinking man, alone of all creatures, managed during a brief history to create the world-wide energy and raw material crisis? For some 2,500 million years there has been life on our planet; the higher plants have existed for about 450 million years, vertebrates for some 400 million. Why did these not produce a similar crisis during the much longer span of their existence? They too transform matter and energy, and on an even greater scale than man has ever done or will ever do. But they have caused no crisis. Why? And can we learn from animals and plants? I think we can – and must, if we are to survive – as the plants' example will demonstrate in this book.

I shall not only show that plants are more 'conscious of the environment' than men, and therefore will never be overtaken by a crisis; I shall also make clear why this is so, and at what points, in his capacity of designing engineer, man has made, and is still making, the gravest blunders. I shall attempt to explain how we can judge whether a design is good, i.e. whether it passes the essential test – does it meet the requirements of ecology, both now and in the future?

In 1859, when Charles Darwin published his *Origin of Species by Natural Selection*, the current conception of teleology in animate nature was at once transformed. Previously, scientists had believed that all plants and animals were exceedingly ingenious 'designs', carefully adapted in every detail to the environment in which they lived. Darwin's evolutionary doctrine proved that there could be no question of design. His selection theory assumes nothing more nor less than that nature produces, without plan or meaning, a great multitude of species and forms, of which only those survive that best withstand environmental conditions. Therefore there can be no question of adaptation, but only of a selection of those species that *appear* to be adapted.

Since Darwin, biologists have been firmly convinced that nature works without plan or meaning, pursuing no aim by the direct road of design. But today we see that this conviction is a fatal error. Why should evolution, exactly as Darwin knew it and described it, be planless and irrational? Do not aircraft design engineers work, at precisely that point where specific calculations and plans give out, according to the same principle of evolution, when they test the serviceability of a great number of statistically determined forms in the wind tunnel, in order to choose the one that functions best? Can we say that there is no process of natural selection when nuclear physicists, through thousands of computer operations, try to find out which materials, in which combinations and with what structural form, are best suited to the building of an atomic reactor? They also practise no designed adaptation, but work by the principle of selection. But it would never occur to anyone to call their method planless and irrational. The result is satisfactory, but the way that leads to it is not design but development.

Characteristically, biologists and technicians understand, under the one word 'evolution', two completely different concepts. For the former, 'evolution' is the interplay of planlessness and survival chances. For the latter, it is development, the legitimate means to an end – the best means even, for designs must first stand the test of practice, but the products of development have already stood the test by developing. Designs can be mistaken; developments cannot.

Here we have another great distinction between development and design: developments follow on environmental conditions, as though they could not progress more quickly than these conditions. Their products will always be adapted to the environment. They will not outstrip it and therefore force it to adapt itself to them. Designs have other time standards. They can be speeded up almost at will. That is the root of the evil. Obviously, fatal consequences always supervene when designs outstrip the environment and leave it no time to adapt itself to them. Rivers which are contaminated by industrial waste more quickly than their self-cleansing process can regenerate them provide an example of this. Thus the machine does not belong to the environment of *natural* man; rather, natural man belongs to the *biological* environment of the *artificial* machine. The dilemma of our time consists in man having to adapt himself to the machine, because its maturing principle, which is design, advances more rapidly than the maturing principle of man, which is development. Hence the progress of the machine may prove fatal for man. And since he himself is its design engineer, he is committing suicide.

Man is and remains a biological organism. He cannot be remade overnight. And a good thing too, for that cuts out the danger of mistaken designs. Furthermore, he is forced to take himself into consideration in his all too audacious technical designs, and not to outstrip himself. If he does, he will be as little able to adapt himself to his own products as is nature in general.

When birds, in remote ages, learned to fly, they conquered the air through development, and in the end they became adapted to it. A biological organism needs time for that to happen. But a land animal that I carry up into the air in an aeroplane and then throw out obviously cannot adapt itself to the quickly changing conditions, and hurtles to the ground.

Development leads to adapted forms, but in the case of design this is not so. Why then do we still believe that Darwinian natural selection has nothing to do with intelligent adaptation, but that specific design has? We shall have to get rid of this belief as quickly as possible if we wish to acquire the art of survival. In future our engineers will have to develop rather than design. At every step they take, they will have to look to the compatibility of their artefacts with the natural environment. This does not mean that the specifically human mode of progress, which is design, has to be destroyed root and branch. But where we employ it, we shall have to do so within the two limitations that are taken for granted in development: it must be carried out with a continual feedback to the environment, and its speed must not outstrip the rate of adaptation possible to the natural environment, especially that of man.

As neo-Darwinians let us cease, in reading this book, to believe that plants are the products of chance, organisms which are not adapted to their environment but continue to exist only because they have been lucky enough to meet its demands. It is simply not true that development and adaptation are irreconcilable with one another. Development furnishes adapted forms, as we shall see at every step – forms adapted in the best way possible. That is why plants are better engineers than we are.

CHAPTER TWO

The Power of Light

The most powerful source of energy, and at the same time the one least effectively exploited by man, is light. Except for atomic energy, practically all the energy used by man comes, indirectly, from sunlight. Hydroelectric power plants use the downward flow of water. Without the water cycle produced by evaporation they would dry up. Evaporation and rain, however, are a result of the sun's heat. Thermal power stations work with coal, oil or gas – organic substances that plants have stored up with the help of solar energy over millions of years. Petrol, diesel oil, kerosene, petroleum, natural gas and other heating and power fuels have the same origin. The foods by which we give energy to our bodies (often more than is advisable) may ultimately be traced back to plants, for the animals we eat are dependent on plants or on other plant-eaters for their nourishment. And the plants build up their substance directly with the help of the energy they get from light.

Without this primary use of light no life would be conceivable. Plants are capable of utilizing energy from light; man with all his technology has been able to do so only during the last few decades, and on a scale which can be safely ignored even today in any normal consideration of energy production. The ability to build firm organic substances from air and water with the help of light is given only to the green plant. In the course of their assimilation in leaves and stalks, carbon dioxide and water somehow give rise, through the action of chlorophyll, first to glucose and then to starch and more complex compounds, and the energy contained in these substances is taken direct from sunlight. These facts are today a commonplace to every schoolboy, and a puzzle to every scientist. No one so far has been able to explain how this simple, yet for man inimitable, chemical process is brought about. All we can be sure of is that it *is* brought about, and that it is *only* brought about when sunlight is available as energy. If I partially cover an assimilating leaf with a small patch, that leaf will produce glucose, and thence starch, from air and water in exactly that area which is reached

by the light. If, after a sunny day, I take a leaf covered in this   1
way and treat it with an iodine solution, the areas containing
starch will go black, and I shall get, as in a photographic
negative, a faithful rendering of the covering mask.

Assimilation is taken so much for granted that the uniqueness
and extraordinary significance of this use of energy is no longer
perceived. Let us examine a few figures which reveal what is
really behind the word 'assimilation'.

We can begin in a small way. A square centimetre of the
leaf of a sugar-beet can synthesize every day only the thousandth
part of a gram of glucose from air and water, with the help of
light. Figures for barley and potato leaves are slightly lower
still. Relative to the entire leaf surface of a plant, however, such
productivity is not to be despised. The amount of solid matter
originating from air and water in the course of a single year is
brought home to us by the yield of a single cornfield. From the
point of view of energy balance, the productivity is also extremely economical: for the synthesis of 1 kilogram of glucose
the green plant needs only about 4·4 kilowatt-hours – the
power consumed by a colour television set in fifteen hours of
viewing. That kilogram of glucose contains 400 grams of pure
carbon, extracted from 0·75 cubic metres of pure carbon
dioxide. Because this gas is contained in the air in a proportion
of only 0·033 per cent, the carbon dioxide has to be extracted
from 2,250 cubic metres of air. At the same time 0·6 of a litre
(just over a pint) of water has to be chemically decomposed for
0·75 of a cubic metre of oxygen to be produced.

Let us now consider assimilation on a world scale. Every year
terrestrial plants store 17,200 million tons of carbon, and
marine plants as much as 25 billion tons, from the carbon
dioxide of air and water. This total of 42,200 million tons of
carbon is contained in 105,500 million tons of glucose, which
corresponds to a goods train more than 30 million miles long
filled to the brim with glucose. It would be long enough to
cover the entire railway network of the world 40 times without
a break; it would be 130 times as long as the distance from the
earth to the moon, or a third of the distance from the earth to
the sun. This train would contain the glucose production of only
one single year.

In the course of assimilation 467 billion kilowatt-hours of light
output are absorbed annually. That is over a hundred times the
entire world production of electric power (according to figures

for the year 1973). Of this the plants of the earth use 189,500 billion kilowatt-hours annually for their own energy needs. 17,100 million tons of carbon return into the atmosphere. But 25,100 million tons of carbon remain stored in solid matter, in addition to 37,800 million cubic metres of water. 277,500 billion kilowatt-hours are sunk every year in that storehouse of energy, the plant. For this giant-scale synthesis, 79,000 billion cubic metres of carbon dioxide are annually converted by the plants. 32,000 billions find their way back into the air, 47,000 billions remain bound. But on the other hand, 47,000 billion cubic metres of pure oxygen are precipitated.

The figures are incredible. And the balance as it stands would soon mean a scarcity of carbon dioxide (and thus the starvation of all life on earth) were it not that men, animals and micro-organisms are continually burning up the carbon they absorb with their food, and turning it into carbon dioxide which they breathe out. We human beings alone annually return 140 million tons of carbon to the atmosphere when we breathe, while animals and soil bacteria return over 24 billion tons. The smallest of vegetable organisms, bacteria, are of the utmost importance in this process. Without that return of carbon to the atmosphere, terrestrial plants would die of starvation in 62 years, though marine plants would have provision for 2,400 years. By and large, therefore, through the co-operation of all living things, the balance of carbon is adjusted. For millions of years there was a slight surplus in the form of stored carbon: peat, coal, petroleum and natural gas. Since the end of the last century man has been systematically using up these deposits. It has been calculated that between 1890 and 1960 an annual average of 1,200 million tons of carbon was returned to the atmosphere through industrial combustion processes. Thus those 70 years account for more than 13 per cent of the atmospheric total. In the year 1960 the influx of carbon from industrial plant amounted to 3,120 million tons; since then it has rapidly risen, and is going to rise still more in the future.

The American oceanographers Roger Revelle and Hans E. Suess have expressed it thus:

Humanity is busy instituting a geophysical experiment in the grand style, one that could not have taken place in the past, and one which it will not be possible to repeat. In the course of a few centuries we have been returning to the atmosphere and the oceans the concentrated sedimentary carbon of millions of years.

We are busy upsetting the atmospheric equilibrium. Will this be crucial for the future development of climate and life on our planet? Our need of energy, which grows from year to year, has hitherto made this extremely dubious experiment necessary for, unlike the plant, we do not yet know how to make direct large-scale use of sunlight.

That the opening up of that cheap source of energy, light, does not necessarily presuppose costly projects of research such as the installation of giant sun-furnaces (on which work is now being done in the USA) is proved by a quite recent chemical discovery. Many plastics which up to now have been piling up intact on rubbish tips, or producing highly poisonous gases in incinerators, are quickly broken down by sunlight without cost or danger to the environment, if an enzyme required to catalyse this reaction is mixed with them in their manufacture.

If the energy of sunlight is to be used, the important thing is to catch as much light as possible. With space probes, satellites and moon-landing bodies, especially in the Skylab project, the problem is very real. The electronic measuring instruments and transmitters of artificial space stations are frequently fed from so-called solar batteries. These are cells which, as in a photographic exposure meter, convert sunlight directly into electricity. Hence the astronauts do not need to carry 'canned energy', such as heavy storage batteries with a limited life.

If firmly fixed to a moving system such as a satellite, the solar cells would be illuminated by the sun only intermittently, and most of the time would lie inactive in shadow. To avoid this loss of output, space technicians have developed highly complex electronic tracking systems, which sense the direction of the sun's beams, and then, by means of control motors, move the solar cell panels to face the sun in order to receive the maximum amount of light. From the simple measuring technique (of the direction of the beam), through the logical interpretation of the data, to the precise execution of control, is technically a long step; and the electronic-mechanical construction which can do all that is correspondingly expensive, not only in its operation, but also in its manufacture. In addition, the circuits connecting sun sensors through the computing system to the control motors occupy considerable space.

As with most intricate problems, plants are here far superior to our technology. They are masters of phototropism, which is the name given by botanists to the movement of a plant under the stimulus of light (causing indoor plants, for instance, to turn towards the window); but how small are their means, and how minimal their expense! Apparatus for the measurement of the direction of light, interpretation mechanism, control propulsion, and finally the part itself that has to be controlled, form one compact unit – in the extreme case of unicellular bacteria, a thousandth part of a millimetre in size. In higher plants, too, the phototropic mechanism is contrived with such economy that it actually takes up no space of its own at all. It is only one of the many ancillary functions of the whole plant.

What does phototropism do? In space-travel it orientates an instrument according to the position of the sun; in botany, it performs such a multiplicity of separate functions that study of the subject is far from being exhausted. Here we shall examine only a few examples to see how plants react externally to the stimulus of light.

First of all, light plays a decisive part in growth. That a long, thin, almost leafless, yellow-white shoot issuing from a potato eye in a dark cellar is different from a healthy, broad-leaved dark-green potato plant in the field needs no demonstration. The meaning of such a marked difference is obvious: whereas the shoot that gets the light can draw on the sun's energy for its progress, the one without light must draw on the reserves in the potato tuber. Through its long thin shoots it is trying to get up to the light as quickly as possible, and with the minimum expenditure of energy.

Another effect of light on plant form produces exactly the opposite result. Too much light, especially that of the rich ultra-violet rays of mountain altitudes, can be harmful to man and lead to sunburn. Plants which receive intense light protect themselves against it, e.g. the edelweiss with its silvery-white fur. In flat country, where the flower receives much weaker rays with a very small proportion of ultra-violet, hair-formation is immediately reduced, and where lighting conditions are bad, hair is totally absent.

Light exercises direct control on the growth of plants. They grow towards the light, and in such a sensitive way that in many species a shoot that has been kept a whole day in the dark will react to a single flash of light lasting no more than two-thousandths of a second. After about twenty minutes the reaction sets in, reaching its peak in about an hour's time. The extent of the bend described by the plant will depend directly on the amount of light that has fallen on it.

Leaves normally place themselves at right angles to the direction from which light falls. (Tropical foliage is an exception; it must protect itself from excessively strong rays.) If one leaf is overshadowed by another, it will bend sideways so as to receive as much radiation as possible. Over a whole tree or bush the control of each leaf individually results in a veritable leaf-mosaic, in which leaf elbows leaf, and advantage is taken of every gap. Ivy-covered surfaces show this to perfection, and here the shape of the leaf contributes to the mosaic appearance.

The toadflax (*Linaria cymbelaria*) possesses a particularly ingenious phototropic system. The stalk reacts in such a way as to turn the flower to the light. Later, when the flower withers and is replaced by the ripening fruit, the stalk will suddenly behave in exactly the opposite way: it will turn the fruit from the light and seek dark clefts in a wall or rock in which the seeds can be sown.

If one were to take tender plants from tropical forests, and try to grow them in the open air of a botanic garden in temperate Europe or North America, the immigrants would not last long. The climatic differences between the constantly moist, luke-warm soil of the primeval tropical forest and the environmental conditions of the so-called temperate latitudes are obvious. For instance, if the average annual temperature of Java is about 25°C, the average temperature of the coldest month, February, is about 24·5°C, and that of the warmest month, September, is about 25·5°C. By contrast, the average temperature for January in Munich is −1·5°C, and that for July 17·5°C. Thus in the tropical forest there is a difference of only 1° between the temperatures of the coldest and the warmest months, but in the 'temperate' climate there is a difference of nearly 20°. This variation is true also of the distribution of humidity, of rainfall and of sunshine during the year.

Not being geared to the radical weather variations of the 'temperate' climate, tropical plants need the protection of a greenhouse to provide more favourable temperatures, greater atmospheric humidity, and more moderate light than are usual in the open air of the 'temperate' zone. But there are climatic regions with more extreme conditions still. In some deserts, temperatures can vary overnight by 50°C or even more.

Relative humidity can vary in the same space of time by 70 or 80 per cent; added to all this are the unremitting rays of a merciless sun.

In a climate of such extremes as in Namaqualand in the south of Africa, there is one plant which is quite at home, for it has contrived to protect itself by a peculiar kind of greenhouse: it

3   is the 'window plant', *Fenestraria*, which lies almost completely buried in the ground with only the extreme ends of its thick, club-like leaves protruding. Through its underground life this open-air plant becomes practically an indoor variety, for the soil round about protects it from excessive heat, from harsh light, and from dehydration. But every green plant needs some light to live, not a great quantity concentrated in only one small area, but bright, though mild, light, distributed as evenly as possible. The form of growth in the window plant seems to belong to the first, undesirable alternative. Almost the entire body of the plant lies underground in the dark; only the ends of the leaves reach above the ground and receive the blazing sun. But it is here that the greenhouse trick comes into play. The ends of the leaves have no chlorophyll. They are absolutely transparent, like glass. The light does not burn them, but passes through, practically undiminished. The inside of the fleshy little leaves consists likewise entirely of a tissue clear as

4   glass.

Through the watery tissue the sun's rays finally reach the subterranean leaf walls. These alone are not transparent: they contain the green colouring substance, the chlorophyll, by means of which they can make use of the energy of light. But the light that reaches them has been repeatedly reduced. First, the 'window' at the extremity of the leaf wards off a great part of the ultra-violet rays, exactly like a greenhouse window; then, by dispersion inside the leaf, the light which falls through the small surface of the 'window' is fairly evenly distributed over the large surface of the leaf wall; and finally the glassy matter also absorbs some light, just as a window-pane does. All in all, therefore, *Fenestraria* applies the greenhouse principle exactly, and yet, though the result is the same, there is a fundamental difference between greenhouse and plant. The greenhouse is meant to adapt environmental conditions to the necessities of exotic plants. The *Fenestraria* adapts its necessities by the same means to the environment. The one is a predesigned alteration *of* the environment, the other an adaptation *to* the environment.

CHAPTER THREE

The Waste Problem

We have all heard reports of cities in which the refuse collectors have gone on strike. Dustbins overflow, and the garbage they can no longer contain strews the streets. Every gust of wind sweeps old newspapers, soiled cartons and rattling tins in front of it. Greasy dust whirls around, and the air is laden with the nauseous stench of decayed food.

Such is the rubbish problem on a small scale. On a large scale it begins with the accumulated refuse not of a few days, but of months, years, decades. Think how many ordinary appurtenances there are in daily life, from the ball-point refill to the motor-car, or how many industrial products roll off the automated assembly lines, second by second. Without exception, every single thing that is produced for use is potential rubbish. Tomorrow or the day after these things will either be used up or simply discarded. Refrigerators, prams, files, used engine oil, obsolete armoured cars, entire houses – literally everything produced yesterday is the refuse of today, and everything manufactured today will be the refuse of tomorrow. That is as certain as the alternation of harvest and sowing. Why then do we not think of the problem of disposal at the manufacturing stage, since it is a problem which we are bound to face?

Iron rusts; but the durable goods of our economy must not be allowed to rust. So we prefer to make them out of aluminium rather than iron. This is quite reasonable: higher production costs are offset by greater durability. But why do our packing industries produce more and more cans, for which sheet steel was formerly used, out of aluminium? The production of aluminium requires six times as much energy as does the production of iron. It is true that aluminium cans are lighter than steel ones, but twice as much energy goes into their manufacture. Double expenditure of energy not only doubles wastage of energy reserves, it also doubles the pollution. In no way does the product justify the increased pollution or the expense involved in its manufacture. Cans have a very short life in our economy, but aluminium cans live almost for ever on the garbage dump.

The refuse production of plants, in contrast with that of men, has been going on for millions of years. But disposal takes place silently, without offensive odours, without harmful exhausts, and without the pollution of land or water. Plant refuse does not lie unconsumed for years, as do old plastic containers,

5   abandoned cars or air-raid shelters. It is swiftly and continually broken down and made usable again. Production and decomposition cancel one another out. A system so well balanced is capable of functioning for an unlimited time.

Man has never really observed or understood this ebb and flow in nature. If he had he could scarcely have brought himself to create anything so unbalanced as our technology. Up to the present there has been no appreciable degree of destruction to match production. Even what is called the utilization of refuse scarcely deserves the name. Combustion equipment merely transforms solid into gaseous filth, and water-handling apparatus generally turns liquid into solid refuse. The root of the evil lies much deeper. If we do not stop producing things ceaselessly without giving any thought to the problem of their ultimate disposal (not in the sense of 'laying aside' but of 'making disappear'), no programme for the protection of the environment, however costly, will succeed in solving the world-wide refuse problem. The fact that our language has up till now lacked even an approximate term for the process contrary to production, that of reducing manufactured goods to their original materials, shows how one-sided our production thinking is, how completely devoid of a sense of building up and breaking down.

Plants, on the other hand, have developed the most perfect balance imaginable between production and 'deproduction': a total recycling or complete circulation of material. This means that everything plants no longer use is immediately decomposed, reduced to its original materials and at once used again. There is really no sharp distinction in botany between coming into being and passing away. Building up and breaking down merge with one another. The decomposition of leaves, stalks and blossoms that are no longer required, and the production of new plant organisms, go hand in hand. Every substance produced is developed from the beginning in such a way that it can at once be simply, quickly and above all usefully decomposed. The smoothness of this transition from becoming to passing away is most impressively demonstrated by the career of a series of evergreen tropical trees and bushes. Here, since the old foliage does not fall away till long after new leaves have developed, it makes way for the young and more efficient: its leaf-stalks bend down sharply, so that the ageing leaves do not overshadow the young ones, and hinder their growth as little as possible. The old leaves do not go limp or wither, but move quite actively into the position favourable to the younger generation.

Before the leaves fall, as they do in the end, they wither on the plant or change colour. And it is exactly the same with our own native foliage. This is the outward sign of the extraction by the plants of important nitrogen compounds from the leaves they no longer require, before the leaves fall off. These compounds find their way back into the branches, where, transformed, they are once more employed as building material. Thus plants extract usable substances from their refuse before they throw it away.

The leaves that have been shed cover the floor of forest and woodland with a thick layer in winter and early spring, not by any means as useless refuse, but as something which encourages new growth. Just as the gardener covers his young plants with brushwood and mats to cut down heat loss and to protect them from cold winds, so the fallen leaves protect the ground vegetation of the forest. If new growth stirs in the early months of the year, the gardener uncovers his beds. But nature cannot roll up the carpet of leaves. Nor, from one point of view, would that be desirable, since the leaves keep the ground damp and warm, exactly as the germinating seed requires. But the young plants need light for their growth. The leaves of the previous year behave almost as if they were aware of this necessity: they become transparent to light rays, especially in that part of the spectrum that is important for photosynthesis, and reach their greatest transparency exactly at the time of germination, from March to April. Fallen leaves, therefore, are a form of refuse which could not possibly be more favourable to the environment. Only when they have fulfilled their last function, the protection of the seedlings against loss of humidity, are they at last broken down by ground bacteria, thereby serving as nourishment for these, the smallest of all plants. What remains is fertile humus.

Plants are indeed inspired refuse exploiters, and not only in relation to their own material. Animal refuse also – excrement and carrion – is swiftly and completely processed by bacteria. Graphic examples of the multiple uses of bird droppings are provided by plants growing high in the branches of a tree. There are many such plants in tropical regions all over the globe. If their seeds are not adapted so as to be carried by the wind, they are packed into soft, sweet fruit pulp, and rely on transport by birds. The berries are eaten, but the hard seeds within cannot be digested by the bird's stomach. Thus they fall with the droppings. If the little seeds were laid by themselves on the frequently smooth tree-trunk, they would simply fall down. The bird-dung gives them solidity and affords them, at the same time, the moisture necessary for germination. As though this were not enough, the young growing plant can use it as a welcome manure, rich in nitrogen – an ingenious use of refuse material.

Refuse, it will be seen, can be extremely valuable, occasionally even vital. The international magazine *Plain Truth*, which concerns itself with the problems of everyday living, recently published an article which showed how the little town of Santee on the West Coast of the USA had solved its sewage problem. The town authorities realized that what was required was the complete disposal of refuse through treatment and recycling. Under the heading 'Sewage, a vital raw material', the author writes:

Complete sewage treatment – whether by natural decay in the soil or by man's technological processes – produces clean water and nutrients for plant growth as end products. Whereas nature recycles wastes in the soil, the Santee process uses technological means to separate suspended and dissolved wastes from water.

Santee planners initially wondered: 'Can waste water be reclaimed on a large scale?' As facts were gathered they realized that sewage is not a totally undesirable substance. In reality it contains valuable raw materials.

Santee leaders envisioned the creation of marketable products from waste water. Sewage solids make high-grade soil-conditioners and plant fertilizers valuable to gardeners, horticulturists and farmers, as long as the sewage is free from heavy metals and other industrial wastes. (These pollutants escape removal by even the most advanced sewage treatment facilities, so that waste water containing non-biological materials must be treated separately.)

Pinched between biological necessity and sewage disposal cost, the vast majority of modern cities give their waste water a 'once-over' treatment, then pipe the effluent to the nearest stream, river or large body of water for dilution. This method of sewage disposal, is called 'primary' treatment. Most cities throughout the world use only primary sewage treatment, if any at all. Of course, primary treatment is the quickest, least expensive – and one of the least satisfactory – means of ridding a community of its waste.

For little thought is given by most communities to considering the effects of their actions on others. Humans seldom apply Jesus' Golden Rule: 'do unto others as you would have them do unto you', in regard to mundane matters such as sewage disposal. If they did they would never think of dumping wastes upon others, either individually or collectively. Unfortunately man too often takes the modified escape route of: 'do unto others what they do unto you – only *do it first*!'

And today the same attitude is exhibited nearly everywhere. New York City, for example, that overcrowded town with its millions of households and giant industrial plants, dumps its waste water by way of the Hudson into the sea. The mighty Mississippi, carrying untreated sewage and industrial refuse by the ton, streams stinking into the Gulf of Mexico. Nor are European waters much cleaner, in spite of the campaigns against pollution of the environment.

Whether one lives in the abundance of the western world, or in the poverty of the third world, the pollution of the environment is a fact with which both there and here one has to live, though the situation might be changed.

Why is the Santee project so important?

First, it puts a stop to water-pollution, and thus saves many downstream dwellers from a deterioration of their water-supply and injury to health. Second, it supplies purified, fresh water, which becomes every day more important, since clean water is increasingly scarce.

Today the Santee project has proved itself to be a model for modern sewage treatment plants. No pollution results from the process, and some eight million litres of raw sewage are processed every day. Reclaimed water is available for many uses.

Representatives from over forty countries have visited the project. Indeed it is a sensible model for any town attempting to stretch its water supply while solving a major cause of pollution. The Santee project pictures the kind of resource-consciousness necessary for a clean world of the future.

What is lauded here as an exception is the usual thing in the world of botany. There all natural pollution of waters is naturally eliminated. Foreign substances are treated biologically by water-plants and bacteria and fed back into the ground. And as a perfect model for the immediate treatment and re-use of domestic and industrial water, consider the urn-like leaves, traversed by roots, of the tropical liana *Dischidia rafflesiana*, which will be discussed in Chapter 8.

CHAPTER FOUR

Plants as Chemists

If two parts of hydrogen are mixed with one of oxygen, the result is oxyhydrogen gas. If a match is put to the mixture, it will explode, the two gases combining chemically as water. But the reaction of the two gases to one another can be produced without igniting them, simply by adding platinum black. The platinum will cause the hydrogen and oxygen to combine at room temperature, as water, giving forth great heat. The platinum itself remains unchanged. But its presence is enough to start the reaction, just as the mere presence of a policeman can compel motorists to pay attention to a speed limit.

Such mediating substances are called catalysts. Chemists often use them to speed up processes in the laboratory or in industrial manufacture, or to make them possible on a larger scale. Platinum is only one of a whole range of such catalysts, without which the technological chemistry of our epoch would not exist. Cigarette ash can act as a catalyst; lump sugar sprinkled with it can be ignited and will burn with a blue flame. Without the cigarette ash the sugar could not be ignited. A certain quantity of hydrogen peroxide (H_2O_2), when heated, will disintegrate into hydrogen and oxygen in time, without a catalyst. If platinum black is added, however, the operation is a thousand times quicker. Chemical processes thus become vastly more economical through the use of catalysts.

Chemical reactions are taking place every moment in every living plant. And the plant, too, uses a whole range of highly effective agents, most of which, unlike the chemist's catalysts, it has developed for a specific reaction. Their catalytic potential is therefore a great improvement on that of the substances which we know in the laboratory and in industry. If the time needed for the disintegration of hydrogen peroxide is reduced by the technological auxiliary platinum to a thousandth, the effect of the vegetable substance *catalase* is a thousand times better – for it reduces reaction time to a millionth.

I need not stress that the plant is in many ways superior to the chemist: one need only think of its ability to build up glucose and starch out of water and air. But the case of the catalyst demonstrates that even when the plant is doing something in no way different from what the chemist does, it is infinitely his superior. Here once again its principle is development, corresponding to the environment and for that reason in harmony with it. Through the meticulous adaptation of the catalyst to a given chemical task its effect is enormously increased, and thereby its profitability.

Fruit growers, in the course of storing and dispatching apples, have made a remarkable discovery. If late varieties are picked before they are ripe and packed for transport they will ripen in a given time. But if they are packed together with early-maturing varieties, they ripen considerably faster. They are therefore obviously stimulated to early ripeness by their forward companions. And for this to happen the two kinds do not even need to touch one another. How is this mutual influence possible?

Many growers of indoor plants will have noticed that different specimens of the same species will bloom on the same day; indeed, buds emerging later on one plant will overtake more developed ones on another, so as to burst open with them at the same time. This is certainly practical, for simultaneous blooming enables them to be simultaneously pollinated by insects. But how can plants standing in separate pots come to an understanding about progress in bud development?

On dry steppes and in semi-deserts the fight for water is a fight for life or death. There is enough space for a thick carpet of vegetation, but enough moisture in the ground for only a few. Therefore many plants in dry regions wage a regular war with one another. They make life difficult for a neighbour which might draw away precious water from the soil; they hinder its growth and delay or hinder even the germination of its seeds in their vicinity. But what are the invisible weapons used in this warfare?

In the guerrilla war between steppe and desert dwellers, just as in the ripening of apples and the development of buds, every effect from contact can be excluded. Optical, acoustical, electrical or similar methods of communication are out of the question. Influence, therefore, must be of a chemical nature. And because, at least in the case of the apples and the plants in separate pots, there is not even any connecting soil in which the plants could be rooted and thus influence one another, chemicals in liquid form as intermediaries are likewise ruled out. It must therefore be gases that produce these remarkable phenomena. And indeed it has been discovered that ripening apples emit small quantities of ethylene gas, which is capable of strongly influencing other plants. The later varieties of apple ripen in storage under the effect of this gas, and young bean plants, for example, will grow much more slowly than usual in 'apple air', but they are stronger. A multitude of these 'wireless' influences of various plants on one another have been observed. They include not only ethylene gas (which incidentally is faintly present in our gas fires and stoves) but also a whole series of other agents, of which only a few have been scientifically investigated.

In the chemical war carried on by steppe and desert plants, it is the roots that discharge active substances in solution. The advantage is that they cannot be blown away, as gases would be, by the frequent desert winds, and that they extend their influence over a wide area, as desert plants often have extremely long roots. Stringy roots over fifteen yards long and frequently no more than an inch or two below the ground are not uncommon with desert shrubs only one or two feet in height.

Man, too, has discovered the use of chemical weapons, for the art of war has always been somewhat ahead of other fields of human activity. But the use of harmless chemical substances in the smallest possible concentrations for peaceful communication is a matter of sheer utopianism for the chemical technologist. Yet it would be well worth our while to devote more research to the multitude of chemically controlled plant inter-relationships; besides the influencing of growth, these include such phenomena as the promotion of leaf fall, changes in outer shape, contacts between the higher plants and microorganisms in the ground, and certainly many more hitherto quite unknown modes of behaviour.

Unicellular plants such as bacteria, flagellates, algal and fungal spores, or the sex cells of mosses and ferns, are considered to be the most primitive vegetable organisms. 'Primitive', however, is a questionable term, when one consideres the extraordinary capabilities of these microscopic creatures. In Chapter 7 we shall see how they use in their locomotion a motive power that puts all our technical propulsion mechanisms in the shade, at least as far as effectiveness is concerned. But where in fact are unicellular plants going? Are they making for specific goals through definite directions in their swimming, or merely winding their way more or less at random through the waters?

An English behaviourist once declared that every action, every movement in nature (including man's) subserved either eating or sex – a somewhat ironic simplification. For unicellular organisms it would be correct. Once again, it is chemical substances that guide these tiny organisms, and often the stimulus of light as well. In fertilization, the mobile masculine sex cells find their female partners through quite definite chemical lures. The spores of algae and aquatic fungi take the shortest route to where chemical substances indicate the most suitable ground for germination, and bacteria seek, with unerring certainty, the spots with the highest nutriment concentration. One example will show how admirable the orientation performance of these minute creatures is. The mobile and independent sex cell of a fern is attracted by malic acid, and makes for it wherever it 'scents' it. To do this, 0·000,000,028 (28 US *billionths*) of a milligram is enough, a quantity which a chemist with the most sophisticated modern analysing equipment could identify only with the greatest difficulty. While the chemist needs a special apparatus merely to detect malic acid, the 'primitive' unicellulars also possess an extraordinary capacity for distinguishing between different chemicals. They can recognize and single out with absolute certainty oxygen and nitrogen, albumen compounds and substances related to ammonia, and they can even identify different molecules constructed from similar atoms in different ways (isomers).

The fact that many bacteria with a liking for oxygen are greatly superior in detective capacity to industrial analysis equipment has already been exploited by scientists. When it is a question of indicating the tiniest traces of oxygen and even of localizing them, oxygen bacteria are called to the aid of the investigator. Where they accumulate, oxygen is sure to be present.

It is not only unicellular plants that have developed a scent for nutritive substances. The root tips of all higher plants are capable of seeking out valuable materials and avoiding harmful ones. They grow directly towards the greatest concentration of nutrient salts, as can easily be observed if house plants are grown alternately in clay and plastic pots. When repotting it can be seen that the earth pack in the plastic pot has been uniformly penetrated by the roots. In the clay pot, on the other hand, all the roots press towards the wall of the pot, and the fine little suction roots even penetrate the pores of the clay (and have to be torn off in repotting). This is because in clay pots part of the earth moisture always evaporates through the walls of the pot, and the dissolved nutrient salts remain and accumulate in the clay, while the earth itself has been impoverished. Plastic pots are therefore to be preferred for house plants, but since water loss is much smaller in plastic than in clay pots less watering is needed, or else the plants will rot.

A quite different type of chemical search for nutriment occurs with many parasitic seeds. They germinate only where they 'scent' the proximity of their host plant, or (like the seeds of the mistletoe) if by some means they have landed on the branch of a tree that appeals to them as a source of food.

Another method of immediately and correctly appraising food by its chemical qualities has been developed by the flesh-eating sundew. Its round leaves are covered all over with glandular hairs, on the ends of which are sticky little drops. If an insect flies onto a sundew leaf, it sticks fast and cannot get away. Automatically stimulated, the glandular hairs on the edge of the leaf move towards the insect and add their force to hold it fast. The hairs in the middle, however, do not allow themselves to be disturbed by mere contact. They approach the prey only when they are sure that what has landed is digestible: that is, they immediately analyse the chemical nature of the prey. If albumen is present, or if they detect valuable nitrogenous compounds, they go into action at once, and in addition communicate the stimulus to all the untouched edge-hairs, so that they too bend over the victim. The digestive juices which are then discharged soon disintegrate the food.

CHAPTER FIVE

Plants as Architects

When the first bold structures of steel, glass and concrete began
to oust traditional forms of architecture in the second half of
the last century, a turning-point was reached in the art of
building. New technical paths led to a new aesthetic. Liverpool
Railway Station (1852), the National Library of Paris (1861),
and the Eiffel Tower, erected for the Paris World Exhibition of
1889, are early examples of this development. But its first
brilliant monument was the Crystal Palace, an enormous hall
of iron and glass that broke sharply with all traditional styles.

Its architect, Sir Joseph Paxton, who had been the Duke of
Devonshire's gardener in his youth, entered the competition
announced by the Royal Commission for the design, plan and
execution of a building to house the world exhibition planned
to take place in London in 1851. His ambition to outdo his rivals,
and his lively mind, open to every novelty, made him seek an
epoch-making design. He conceived in his imagination a building which, in spite of its gigantic dimensions, would have
nothing heavy or clumsy about it, but rather would produce an
effect of lightness, indeed of weightlessness. It presupposed a
type of construction which, while using building materials most
sparingly, and inserting glass, glass again and yet more glass,
would be strong enough to satisfy all the requirements of stability.

There was no architectural precedent for such a project, for
there are never models for what is new. True, engineers for
close on a century had been demonstrating the carrying
capacity and constructive advantages of iron for bridges, and
Sir Joseph himself had designed, in glass and iron, the biggest
greenhouse in the world in 1837; but a construction of this type
could not be transposed to the great hall that was planned. The
effect would be too massive and weighty if erected, like a bridge,
on heavy trusses. Paxton envisaged a more airy and graceful
building.

Then the former gardener remembered a botanical structure
which combined minimal expenditure of material with enor-

mous stability and carrying capacity. In his youth he had often admired the immense floating leaves of the royal water-lily, *Victoria amazonica*. These are round, up to six feet in diameter 6 and, in spite of their thinness, are remarkably stable. This stability is achieved by complicated strutting on the underside, 7 with ribs radiating from the centre like wheel-spokes from an axle, and flattening as they extend outwards. And to reduce the distance between them at the leaf-edge, they split up into as many as five forking branches, so that at the edge one rib may have become thirty-two smaller parts. In addition, they are bound to one another by flatter struts.

This was the key to the construction of Sir Joseph Paxton's Crystal Palace: from as few main struttings as possible a series of thinner subordinate struts must branch, bound to one another by many fine little ribs. No more delicately articulated creation could be imagined.

On 15 July 1850 the Royal Commission in London tele-graphed its acceptance. Sir Joseph now knew that his model, the *Victoria amazonica*, had not only enabled him to beat the other competitors, but had opened the way to lasting fame. Over a hundred years later, Paxton's Crystal Palace is seen 8 as the 'now recognizable turning-point that gave a new di-rection to the entire development of architectural history', to quote Konrad Wachsmann (*The Turning-point of Building, Structure and Design*, New York, 1961). The actual conception was not the discovery of Sir Joseph Paxton, but of a tropical water-lily; but it was to Sir Joseph's enduring credit to have noted and appreciated this invention and transposed it to architecture.

When technical construction leads to the same result as natural development, or even when natural development is taken as a model, we can be sure that that construction will be on good lines and suited to the environment.

There are two possibilities in the strutting of thin sheets with a large surface, such as are displayed by many leaves. One of them, that of ribbing, has just been demonstrated. This is particularly suitable for aquatic plants, like the royal water-lily, for in their case it does not matter so much that the added structure of the ribs makes the leaves heavier. The water on which they float bears them up.

Very different is the case of large leaves of terrestrial plants, especially in the tropics, where they are exposed to storms and downpours.

The fan-shaped leaves of some palm-trees grow to a length of 5, 10, or even 15 yards, and can be as much as 3 to 4 yards across, giving a total surface ranging from 15 to 60 square yards. Even with this size they must obviously be as light as possible, so as not to weigh down the leaf-stalk, which often has the additional task of moving the leaves into different positions. Not only must the stalk carry the weight of these giant leaves; it must also be in a position to withstand all the forces that come to bear on them. In the Malaysian islands, for example, violent tornadoes accompany tropical thunderstorms that rage nearly every afternoon, pulling and tearing at the surfaces of the leaves. And then there are the torrential rains, during which more water pours down in a few hours than many places in Europe or America would average in a whole month. Large tropical leaves, therefore, must not only be light in proportion to their dimensions, but also extremely stable, in order to stand up to all the hazards of the weather. The structural combination of lightness with stability is a difficult technical problem. Plants have solved it by means of the *corrugated iron principle*.

Even thin sheets of metal become very rigid if they are corrugated or folded zig-zag. The gain in stability can be tested by a small experiment. Take an ordinary thin sheet of typing paper, size A4 (21 ×29·7 centimetres), and fold it longitudinally at intervals of about half an inch, first one way then the other, till you have a kind of corrugated paper. Whereas an unfolded sheet of paper, when placed like a bridge to lie on its two narrow sides unsupported in the middle, will bend by its own weight alone, the corrugated paper in the same position will remain absolutely firm. You can also add considerably to its weight without it bending. A sheet of paper folded in this way, with each end resting on a wine-glass, will
9   easily support another, full wine-glass weighing 8 oz. and placed on the middle of the paper's 9-inch span. Indeed, the weight could be increased (using a container with a square base and a side $3\frac{1}{2}$ inches long) to $1\frac{1}{2}$ lb before the sheet would bend. Thus stability is increased more than a hundredfold, without any additional supporting structure.
10   Nature has made use of this simple method by developing leaves with a zig-zag cross-section. As early as 1873 the botanist, artist, tropical traveller and distinguished naturalist, Professor

G. Haberlandt described, in *Eine botanische Tropenreise*, such self-diminishing leaves:

If anyone wanted to write a treatise on inefficient arrangements in the plant world, he could make use of the giant leaves of the banana plantain (*Muse sapientum*), cut by wind and rain into many strips, as a striking example. But the example would be ill-chosen, as we would see on closer examination. The leaves, the edges of which are in no way reinforced, are in any case easily torn, parallel to the secondary leaf-veins and often right up to the strong middle rib of the leaf. But the edges of the tears soon scar over and the single leaf-strips, which now hang limply down, go on functioning undisturbed. The over-sized intact leaf is thus transformed by the wind into a pinnate leaf. The plant has economized in material, for if it protected so large a surface from rents, bast cords of corresponding thickness would have been necessary. It has at the same time procured another advantage: the hanging leaf-strips, which are thin and not very firmly built, now find themselves in a position that protects them against further damage from heavy downpours, and exposed to the burning rays of the tropical sun overhead at much more acute angles than when the leaf-spread was intact and entire. So the leaf of the banana plantain, tattered by wind and rain, is an instructive example of how, in the realm of organisms, the useful can develop, not only from useless, but even from inauspicious beginnings; and also of how foreign to nature in her phenomena of adaptation would be an obstinate clinging to preconceived schemes, however well-tried.

The stiffening of material by folding is a frequently applied technological principle. Roofs, garages, aircraft fuselages during and after the Second World War, as in the Junkers JU types, even the coachwork of cars in corrugated or pleated sheet metal, corrugated cardboard for packing, balcony facings from corrugated asbestos, cement and polyester sheets, and many other things, down to the pleated paper robes of gold and silver Christmas angels, owe their stability to this ingeniously simple method. But the idea of serrating the structure of the folds from the edge inwards, of cutting it up into single ribs of one fold, as the palm does, has only recently suggested itself to the technologists. In 1965 it was applied to the cantilever roof of the entry into what is so far the longest, deepest and most modern vehicular tunnel in the world, the Mont Blanc tunnel. 10 11

A building element important in the plant world, and for many ages past also in technology, is the column. But while for 4000 years architects have been erecting columns of a completely homogenous construction, identical throughout, nature from time immemorial has grown columns as ingeniously constructed as the ferro-concrete piers known to man only for the last 100 years.

Concrete is strongly pressure-proof, yet it has poor resistance to traction stresses. But at the same time it is sensitive to bending stresses. For example, if a concrete slab lies open-ended between two supports and receives a weight in the middle, like the corrugated paper in our experiment, its underside will stretch and eventually split, for concrete is not elastic. The slab can be reinforced by pouring in with the concrete a steel mesh which resists the stress. Such ferro-concrete structures have quite remarkable strength, as is proved by wide-spanned motorway bridges. Naturally the steel insertion must be exactly at the spot where traction tension occurs; in the case of a slab resting open-ended between two supports, this will be on the underside. In the case of a cantilever balcony, however, the steel mesh must be high up in the concrete because, as the balcony is fixed on one side only, it is inclined to bend in a direction exactly opposite to that of the slab we have just considered.

What happens in the case of a column? As it is completely symmetrical, bending stresses may occur in all directions. It must therefore be reinforced in such a way that the longitudinal rods run through the whole length of the column and all the way round, as near as possible to the surface. And in order that the rod construction should not disintegrate before the pouring, the rods are bound to one another by strong wire. In the interior no steel clamps are necessary, for there no traction tension occurs.

Neither an engineer nor an architect invented reinforced concrete: it was a gardener. The Frenchman, Joseph Monier, in 1867, trying to make plant tubs out of concrete, used iron rods for the first time, in the technique now called after him. Actually he did not really invent reinforced concrete; he discovered it. For every day he noted that plants reinforce their supporting structures in exactly that way. Without that discovery of his, all the audacious concrete structures of our time, bridges, multi-storey buildings, slender television towers, cantilever airport hangars, even our modest petrol-station shelters, would be inconceivable.

Plants have exploited the principle for over 250 million years. The reinforcing 'cages' of plant candelabra can be clearly

seen in dead saguaro cacti, because the rest of their tissue does not turn to wood but frequently disintegrates completely. The skeleton of a giant saguaro cactus stem, like the reinforcement cage of a ferro-concrete column, can be seen to show the Monier   12, 13
process close to the surface. The whole of the interior, both of plant and pillar, is without reinforcement elements. The prickly pear (*Opuntia bigelovii*) displays a differently organized   14
reinforcing tissue. This too lies immediately below the surface of the living plant-body, and there is no reinforcement at the centre.

Not only does the plant arrange its reinforcing tissue in the most effective way possible, but its traction and bending resistance is as strong as that of steel wire. Indeed, its elasticity and extensibility are even superior. Only thus could a wheat stalk $\frac{3}{8}$ of an inch in diameter and 5 feet high bear the heavy burden of the ear at the end, and bend and rise resiliently in the wind.

The reinforcement skeletons of the mighty saguaro cacti of Arizona, 50 feet or more in height, produce, when the plant has died off, an almost grotesque effect. They stand like giants' shaving brushes in the monotonous landscape. In them only the lowest supporting fibres are cross-connected. Further up they   13
stand completely free beside one another in the body of the plant, falling apart when it decays in a fashion reminiscent of a spliced rope or cable-end. Lianas are almost entirely built up of many such single flexible strands, which can change their position relative to one another, exactly like the wire or rope cables of our technology.

But this is not the only way in which lianas have succeeded astonishingly well in mastering the technical problems of their environment.

Nothing is more alien to nature – though not to technology – than an obstinate clinging to predetermined schemes. We have already seen this in the case of the banana plantain, which intentionally tears its own leaves as a precaution against violent tropical storms. The tougher the living conditions, the more ingenious and versatile will be the plant's adaptation to them. Almost always, adaptation is carried so far that the whole outward form of the plant is stamped by the environment. Plants of quite different families, living under the same hostile conditions, often look deceptively alike. In desert regions, for instance, the spherical shape has shown itself to be the most practical, and is especially familar to us in cacti. But not everything spherical and prickly is a cactus. Other plants from quite other families have adapted themselves through this useful shape to a killing desert or steppe climate.

On the other hand, cacti need not always look like thorn-studded balls or like columns. In his book *Wunderwelt Kakteen* the well-known cactus-hunter Curt Backeberg gives the following graphic and impressive report from Cuba of how cacti can look when the environment seems to require it:

An extensive sun-drenched bushland stretches away to the horizon. Out of it low tree-tops emerge, and isolated thick trunks with massy tops. But the boughs look very strange, and they have two different kinds of appendage. Long, silver-grey veils of *Tillandsia usneoides*, Spanish moss, float down from them and wave to and fro in the warm ground-wind, while between them hang numberless strings, thin and interwoven, sometimes moved as if by invisible hands – a few old-established colonies of the epiphytic *Rhipsalis* cacti, looking as though they had fled from increasing ground vegetation to the clear heights of the tree-tops. How multivarious their forms! Here almost thread-like strings, there unwieldy, spread-eagled limbs covered with delicate down, squared chains, artistically serrated, corrugated and dentate runners, and a tangled mosaic of swaying leaves. In blossom time they are prettily garlanded or decorated with tiny coloured spots, and later bejewelled with white, pink, gold, dark-red and blue-black berries.

Cacti like liana-strings hanging from the tops of forest giants are widespread in the tropical jungles of Central and South America. Some are found even in Madagascar and in Ceylon. Creeping and climbing cacti provide an astonishing proof of adaptability to a new environment, yet they represent only one among hundreds. Climbing plants, creeper plants, plants growing high up on other trees, all are numerous in the tropics. They all want to escape as quickly as possible from the constant twilight in the undergrowth of a primeval forest. They seek the direct road upwards to the light, one that needs no preliminary development of substance-consuming stems or support systems under unfavourable living conditions. Therefore they climb up the supports that are already there. They have developed technically perfect organs specifically for this task, and in fact there is no single organ that has not played its part in this development. There are clinging roots, climbing boughs and branches, creeper leaf-stalks or leaf-extensions, grasping floral axes. We find nooses, round turns, adhesive disks with which the plants literally glue themselves to their bases, movable climbing hooks which actively claw the supporting plant and then swell up, retaining clamps and ingenious trap machinery.

G. Haberlandt, in an account of the Botanical Gardens of Buitenzorg, Java (now Bogar), vividly describes the action of the Rotang palm:

If, leaving the footpath, you get into the old liana section you need only have taken a few steps when your hat can be torn from your head, when hooks fasten on to your clothes, and blood flowing from scratches on cheek and hands warns you to take the greatest care. If you look around for the trap that has caught you, you will see that the stalks of the graceful, pinnate leaves of these Rotang palms are provided with flexible and elastic projections 3 to 6 feet long, on which are numerous very strong spines, arranged in a half whorl and looking like backward-curving barbs. Every leaf thus terminates in a terrible whip-lash, which does not easily let go of what it has once gripped. The load-bearing capacity of these 'fishing line' flagella, which consist almost entirely of firm bast tissue, must be colossal. 'One could hang a horse on them,' my guide said in jest, when I tried to calculate their strength. (In several species of *Calamus* it is prolonged inflorescence axes that are transformed into such flagella.) As they are all very flexible, they are very easily slung up on to the boughs of the supporting trees, and their numerous barbs immediately anchor so firmly that no storm could ever detach them. With all its full-grown leaves anchored fast in every direction – spines on the lower parts of the leaf stalks and even on the leaf sheaths frequently add their grip too – the smooth stem goes winding snake-like up through the boughs of the trees, crawls over to neighbouring tree-tops, and finally rises with its youngest leaves above the summit of the supporting tree. Now it can go no farther, as the flagella are whipped around in the empty air. The older leaves, however, gradually die off and are discarded. Bereft of its anchors, the smooth stem slips down under its own weight until the top flagella have once again anchored themselves. The parts of the stem that have sunk down, now nearly as thick as an arm, lie about the ground at the foot of the supporting tree in great loops and intertwining coils, so that it looks as if leafless runners were crawling about looking for other tree-supports. In the Buitenzorg Gardens the longest Rotang stem whose coils can be tracked measures 225 feet. In the virgin forest these mighty ropes are said to reach a length of 600, even 1,000 feet. (*Op. cit.*, pp. 145–6.)

The climbing mechanism of many creeping *Cucurbitaceae* is technically perfect and wellnigh unsurpassable. The co-operation of specialized searching and gripping organs with a sensor and control system represents so brilliant and elegant a solution of a difficult problem that our industrial precision mechanisms and control engineering can offer nothing nearly as effective.

The plant's task is threefold: first it must seek for a suitable hold, then it must get a firm grip of it, and finally it must make sure that the hold will not be broken by stresses brought about by wind or the movements of the supporting plant. Accomplishment is likewise in three stages. First, a systematic search is made to find a hold. This is carried out by a kind of gripping twine that grows perpendicularly, and then bends over to the horizontal, making circular movements like the hands of a clock. In
15 the case of the wax gourd (*Benincasa hispida*), the searching tendrils are only about 6 inches long. But in the tropics there are plants with circulating shoots 3 to 6 feet in length. A tendril of this kind changes its position from hour to hour, seeking as a hold a plate-shaped surface 2 to 4 yards in diameter. Once this is found, it enters at once on the second stage and begins to twine itself round it.

Tendrils of tropical creepers, which are already burdened by their own weight, thicken considerably in the area of the support, thus making the hold firmer and steadier. In addition, the wax gourd ensures that the connection is elastic, to prevent
16 it snapping under a tensile load. The tendril spirals upwards, turning and twisting in the course of its movement.

But the gourd tendril's technical accomplishments do not end there. This string-like structure has developed an acute sense of what affords a firm grip and what does not. Experiments show that smooth supports, such as glass rods, are not grasped; they do not stimulate the tendril, since it cannot secure a lasting hold on them, preferring a support with a rough surface. If it is briefly in contact with such a surface, which is then withdrawn, it will react spontaneously with a groping curve, but soon unbends and tries farther afield. The tremendous technical achievement of these tendrils can only be appreciated by an expert in electrical control engineering and tracking, who would be the first to admit that he could not possibly construct, with so little expenditure, a system even approximately as good.

The wax gourd can do something more: if a tendril finds no

hold it eventually rolls itself up and atrophies. Having no longer anything to do it is dismantled. Tendrils, however, that have got a hold are then strengthened, and even produce wood. It is almost impossible to detach the older ones, which are tremendously stable and at the same time highly elastic through their sprung connections.

This impressive example of *development* in nature is no long-range performance, taking ages for its accomplishment: it enacts itself before our very eyes within a few days. Tendrils grow, seek a hold and, if they do not hit on the right spot, immediately atrophy. The plant then gets valuable material back from them. It is only when they grip a support that more material is invested in them. Nature does not invest in failure.

The lianas' adaptation to their way of life is optimal. If I were to enter into as much detail over all their ingenious devices for survival as I have with the tendrils of the wax gourd, I could fill a book with them. At every step one is aware of a sovereign mastery over the environment, a mastery achieved through adaptation and flexibility in the face of untoward circumstance.

When, more than 4000 years ago, the Neolithic people of western Europe settled on the shores of Lake Constance, Lake Geneva, the Lake of Zurich and the Lake of Neuchâtel, and in the peat bogs of the Lombard plain, they were faced with the problem of building houses in marshy and sometimes flooded ground. How they solved this problem is depicted in old rock paintings and, much later, in the writings of the Greek historian Herodotus: they built on piles.

In 1854 the water level of the Swiss lakes fell very low and well-preserved piles of the old cultures appeared. Students of prehistory now eagerly took up the search for further traces of the lake-dwellers. An idea of what these villages were like is provided by Professor H. Reinerth's reconstruction at Uhldingen on Lake Constance. 17

The fact that the wooden supports of these buildings have survived for 4000 years under water speaks well for the durability of this kind of building. But besides their long life, pile foundations have two further advantages in marshland and regions subject to flooding. They ensure free circulation of air under the cabin, and so prevent the decay of the raised part of the building for a long time; and they lift it high enough above the damp ground or water level to prevent flooding even at high water. This type of construction has proved itself, and is still widespread today in tropical marsh and flood regions. Modern technology also uses it for oil derricks erected on the flat margins of the Caribbean.

Nature has been using this useful building method for many millions of years. Stilt roots give the pandanus and the mangrove, which grow in swamps and by the sea, the same advantages as those enjoyed by the lake villages. Technically, however, the stilt roots are superior to the artificial forms created by man. For while man has to ram his piles laboriously into the ground, mangrove stilts set and sink themselves. This is not done by the roots of the young plants; it is prepared for by the arrangement of blossom and fruit formation in the parent plant. A mangrove seed falling at ebb-tide into muddy flats or shallow water would certainly be washed away at the next high tide, for it would not be able to find any grip on the earth. Therefore, mangrove trees do not drop seeds but, instead, completely developed ramming plugs 24–40 inches in length and of quite considerable weight. Mangroves, so to speak, bring forth their young alive – that is to say, they retain their seeds until they have become viable seedlings already possessing all the preconditions of successful anchorage in the periodically flooded swamp. These plugs, about an inch thick and as smoothly rounded as if turned by a lathe, hang down from many species of mangrove tree, like stout tent pegs almost a yard long. At the lower end they are pointed like a spear; a little way up from the point they become thicker and therefore heavier, so that in falling they remain perpendicular, and when striking the mud penetrate deeply into it.

The young plant has been furnished with food reserves for its journey, so that after such a sudden planting it can equally suddenly begin to grow. Within a few hours it has developed side-roots and managed to anchor itself so effectively in the space of a single ebb-tide that the following high tide cannot dislodge it.

Later, having grown above the flood-level, it develops its stilts, so peculiarly reminiscent of the lake-dwellings, while the main root dies off. In many varieties additional supports grow perpendicularly from the branches, lending further stability to the whole structure.

As so often happens in the plant world, these root supports are better than similar human structures, for, being part of the plant itself, they are more adaptable and more flexible than man-made struts and buttresses. In contrast to design, development is not something settled once and for all: it is alive, and through continuous feedback to the environment remains so. One single big breaker would probably damage a typical lake village quite considerably. But mangroves, which survive with thinner supports, can resist even tidal waves because their stilts are yielding. These elastic roots often grow horizontally at first out of the trunk, and then turn downwards in a wide, sweeping curve. 'Thus the trunk rests on a wide foundation, 2 to 3 metres high, of elastic struts which, as soon as they are struck by a wave, stretch themselves on the extension side, and bend lower on the compression side, finally returning to their original curvature.' (Haberlandt, *op. cit.*, pp. 184–5.)

Heavy constructions covering a relatively small surface area have their own statical requirements. They must be either solidly built or braced and propped like a trellis till the whole structure has attained the necessary stability. The great steel pylons of high-tension transmission lines are a case in point.

For some hundreds of years the same principle applied to house construction, in the form of half-timbering. Walls had no supporting function: they only sheltered from wind and weather. Yet they were still very strong. The last few decades have witnessed a kind of revival of half-timbering. True, we no longer build in wood, and the many diagonal struts have also been dispensed with. Nor do we still speak of half-timbering, but of steel girder construction. But the principle is the same: a trellised framework ensures the necessary stability, while the walls merely exclude cold, rain and wind, and may sometimes be quite thin.

Modern lattice or girder construction will always be more economical than solid building. This is not always immediately apparent in the total cost of the edifice. A steel framework is often dearer than the conventional brick or even concrete, but in the long term there is considerable economy of material, and that is the only thing that counts today, now that we can predict when certain raw materials will have given out.

For thousands of years man has squandered material, and he goes on doing so. Everywhere on earth where there is sandy or gravelly ground, square miles of forest are felled each year, the

19 thin layer of humus removed, and the gravel deposit used as building material. But this ruinous exploitation must cease, or else our children, though they may well live in relatively cheap houses, will be increasingly surrounded by desert and rubbish tips that can no longer be afforested since, together with the forest, we will have also destroyed the soil and the ground water on which vegetation depends.

For the last 200 million years or more, nature has been building with the utmost economy, at a time when there was no hint of any lack of raw materials. The boughs, branches and twigs of trees provide a filigree that fills great spaces with the minimum expenditure of substance. This seems quite natural to us, for we are used to the fact that there are no massive tree-tops. For nature, extreme economy is taken for granted.

Skeleton construction is the answer. Its most striking exemplars are the mighty *Ficus rumphii* trees, close relations of the popular India-rubber tree (*Ficus elastica*) of our living-rooms. The boughs of these wooden giants always grow back towards the centre to meet, penetrate, and mingle with one another. Hence a wide-meshed trellis-work is constructed which gives

20 the tree tremendous stability and enables it to bear its heavy top. Full-grown specimens of the India-rubber tree develop a similar supporting trellis. Here, however, it is not the boughs but a number of closely mingled aerial roots that participate in the skeleton construction.

Other varieties of fig tree skilled in skeleton construction include the epiphytes, which germinate and grow on host support trees. Up in the heights they soon develop a dense system of anchor roots with which they cling to the supporting tree. They then send down the trunk of the host a series of roots which reach the earth and there establish themselves. From the perpendicular nutritional roots, anchor roots branch out horizontally, winding themselves around the supports, and in many places intertwine closely with one another. Thus a strong trellis-work develops that chokes the host tree and kills it. When its trunk later rots, the trellis-like root-grid is strong enough to support the tree-strangler. The stable trellis frame is

12 very like the reinforcement cage of ferro-concrete columns; the only thing missing is the in-filling, which in this case is superfluous.

As we have seen, the skeleton construction method is to be found everywhere in the plant world, as well as in those trees whose branches intertwine to form a supporting trellis, though these are almost entirely confined to the tropics. The fig family has been mentioned in the first place because it displays skeleton construction on a large scale, comparable to the type we are accustomed to in our own buildings. On a miniature scale, the same principle is apparent in every leaf of a dicotyledonous plant. The veins of the leaf always form a network which gives the fragile leaf surface exactly the same support and stability which the girder construction of modern multi-storey buildings give to the thin partitions and outer walls. The 21 fruit of the Chinese lantern (*Physalis alkekengi*) also shows this trellis structure very plainly. 22

Strutting, corrugating, reinforcing, pile and skeleton construction, all of these are methods used in building. As they appear predominantly in relatively large edifices, plants, or parts of plants, it would be difficult to overlook them. But in the area of small – often microscopically small – units the plant works with structures that satisfy all the criteria of efficient lightweight construction.

The sandwich is a comparatively modern invention – not the article of food, which has been with us for a long time, but the layering of building elements to unite light weight with stability. Between two thin, stable supporting sheets is placed a thick layer of light material, usually rigid foam or honeycomb, stuck or welded to the supporting sheets like meat between two pieces of bread. Such constructions are characterized by a remarkable stability obtained with a minimum outlay of material.

The same sandwich structure, which in our technology only became common with modern synthetic materials and light metals, has been from time immemorial the building principle of grass plants. In the wall of a grass stem, the space between the 23 outer and the inner surface is filled with a wide-meshed, very light honeycomb lattice. There could be no more stable structure for a sandwich building element of equally light weight. Regular hexagonal shapes are best for resisting force from without. The bees obviously 'know' that too, for they build their honeycomb with hexagonal cells.

In the modern aircraft industry the same principle, proportionately magnified, is applied to superficially smooth, and at the same time tough, light metal walls. At present the air- 24 craft designer Heinrich Hertel is rendering great service to this industry by investigating certain achievements of plants and animals with scientific thoroughness, and making them the point of departure for new kinds of technical development. The results of mathematically exact analyses of the humming-bird's flight, or of the movements of quick-swimming fish, have been transferred by Hertel to the closely related problems of helicopter rotor blades, for example, or the propulsion of hover-craft. Thus he consciously introduces components of plant and animal development into his construction, and more than once has been able to show into what blind alleys aircraft construc-tion, especially, has been driven by conventional methods.

Sheets of plastic can be reinforced and prevented from tearing by incorporating glass fibres. Researchers have investigated whether all kinds of fibre matting or netting are equally effec-tive as reinforcement, or whether distinctions can be made. If so, what is the ideal fibre structure? The answer is surprising: the thinner a glass fibre, the stronger it is. Of course this does not mean that a thinner fibre is more difficult to tear than a thick one. But it does mean that a fibre of only half a given cross-section can bear *far more* than half the load, before it breaks. For the reinforcement of plastic therefore, glass fibre matting with as many thin, not as few thick, fibres as possible is the best. That is one important discovery. Another is that for the best results single fibres should be 2000 times as long as they are thick. If longer they give no additional strength to the plastic, and in technical production are more difficult to dis-tribute evenly. The practical application of these research re-sults leads, as far as the reinforcement of plastic by glass fibre
25   is concerned, to a quite definite, characteristic structure. This is the result, not of a design, but of a wise development.

Plants keep to the path of development. How did they solve the problem of strengthening their cell walls and making them resistant to strain? The answer is not surprising: the same method – development – leads to the same result. The structure of vegetable tissue is no different from that of plastic rein-
26   forced with glass fibres. For human technology this is a demon-stration of research which is on the right path.

If a medley of short fibres is not strong enough, industry abandons the fibre mat for the fibre net. Simple cross-bar netting has proved efficient. Nature also uses it in cell-wall
27, 28   tissues subject to great strain.

CHAPTER SIX

The Precise Mathematics of Plants

Plants are rational. That explains why certain advantageous principles of construction recur in the most widely varying families. In particular, the most economical use of space is often found in the arrangement of young plant organs when these are developed in considerable numbers. Whether it be leaves on the one stalk, scales on the cones of the pines and firs, single blooms and later seeds in the great round disks of the sunflower, or even the thorny cushions of the cacti, all, without exception, strive from the first moment of their existence for optimal exploitation of space. Just as wine bottles, piled on their sides on a narrow floor-space, will automatically assume a geometrically regular arrangement, so, later, the fully developed plant-organs will be neatly marshalled beside and behind one another.

The ever-recurring yet ever-new picture of distribution in nature has not been without its effect on man. It has so impressed his feeling for form that it will be worth our while to go into this matter, even if it is a little outside the subject of this book, which is the solution of technical problems by plants and men.

Consciously or unconsciously, man derives the principles of aesthetics from his environment. His artistic sense is developed, disciplined and enlivened by everything that meets him in his daily round. From time immemorial we have identified the healthy and the natural with the beautiful and the harmonious. The unnatural, the abnormal, the morbid, is felt to be ugly and discordant. Now if the same principle of construction in plants recurs over and over again, and in a thousand different forms, it is bound to have a lasting effect on our relationship with the world around us.

In 1871 Gustav Theodor Fechner, the scientist and philosopher, presented a series of rectangles to a considerable number 29 of test subjects, and asked them to pick out the one that seemed to them the most beautiful. More than a third chose the rectangle with sides of 21 to 34. This would be a mere statistical

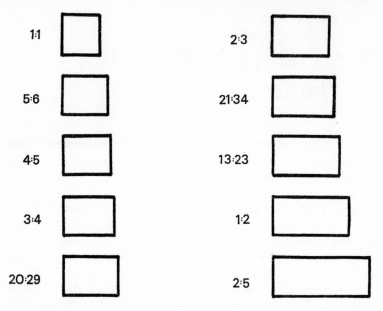

29 Which of these rectangles is the most elegantly proportioned? In Fechner's experiment the second on the right, which happens to have the exact proportions of the so-called golden section, was immediately selected by over a third of the test subjects.

experiment were it not for the fact that plants choose exactly these proportions for their construction plan. Not certainly from any aesthetic feeling, but from pure reasons of expediency.

The proportion 21 to 34 (0·618034: 1) is known to mathematicians and to artists as the golden section or golden ratio. Since the Renaissance, artists have repeatedly and with full consciousness designed their work in accordance with the golden section, which they could observe everywhere in nature, and which had a strange magic for them. Before, this had certainly happened unconsciously. They frequently proceeded from approximate values, such as 3:5 (=0·600) or 5:8 (=0·625). Nature, their model, is far more accurate. The big sunflower, for instance, attains the precise mathematical value of the golden section with an error of only four thousandths of one per cent.

How is it that the golden section manages to express itself in nature? Consider the pincushion cactus (*Mammillaria lanata*). Seen from above it has thorn cushions which start from the vertex in the middle, and travel outwards in spiral lines. This is because new thorn cushions are continually growing out of the vertex, and pushing the older ones towards the edge in accordance with an exact law. Furthermore, there are two groups of spirals superimposed on one another, those that travel clockwise, and those that travel anti-clockwise from the centre. The latter add up to 21, and the former add up to 34 – exactly the proportion of the sides of the rectangle which Fechner's test subjects picked out as the most beautiful, the best aesthetically! Disposition in golden section proportion (0·618034: 1) is here exact to within about 6·5 hundredths of one per cent, for 21:34 =·617647.

The same pincushion cactus looked at from the side reveals 31a that the spirals (over a sufficiently small section of the surface) become straight lines running obliquely from top left to bottom right, and from top right to bottom left. If you draw the grid 31b formed by these straight lines direct from the original, you will see that the lines are at a more acute angle in one direction than in the other; in addition, the inclinations of the lines are so arranged that if one counts them from a point O plotted along the dotted line, exactly 0·618 lines slanting right fall on one slanting left. Naturally no fragments of lines are counted. It then appears that approximately two lines slanting right fall on three lines slanting left (2:3 =0·666), and about three lines slanting right fall on five lines slanting left (5:8 =0·625), and so on. The points of intersection involved recur more exactly on the dotted line, the nearer the number 0·618 is approached.

If the section were wide enough to cover the whole plant, it would emerge that 34 lines inclined to the left would fall on 21 lines inclined to the right, and the last point in the count would be exacly on the initial point O. The line grid thus constituted has exactly the same optimal aesthetic effect as the rectangle proportioned according to the golden section. The lines inclined at definite angles and with different slants give lively yet harmonious tension to the image field. Such tension under the same conditions is to be found in many works of art by old masters, and sometimes also by more recent artists.

Let us, for example, place such a grid over Titian's picture, *Bacchus and Ariadne*. All the important vanishing lines of the 32 picture follow those of the screen. Titian has even drawn a number of less pictorially important forms into the natural field of tension by which his picture is built up: the low mountain to the right near the church tower on the horizon, the branches of the great tree, the edges of the cloud under the constellation of stars, the hind-legs and belly-line of the leopard, the axis of the overturned bars in the foreground to the left, and the lifted right hand of the satyr crowned with vine-leaves to the extreme right of the picture, not to mention the swinging horse's hoof.

Anyone who thinks all that is mere chance, or that Titian's picture is an exception, is advised to trace the grid on transparent paper and place it on the plates of some art publication. The number of times that the composition obeys the golden section dynamics (sometimes also by exact mirror-image) will be remarkable. Statues like Michelangelo's *Libyan Sibyl*, pictures like Tintoretto's *Adoration of the Magi*, Parmigianino's *Madonna of the Long Neck*, Tiepolo's *Asia* on the staircase of the Residenz in Würzburg, Poussin's *Bacchanalian Revel before a Term of Pan*, Brouwer's *In the Tavern* or Watteau's *Love Duet* (mirror-image) are only a few instances.

Artists have always copied, consciously or unconsciously, the model that taught them aesthetics, nature. The simple and effective geometrical solution of natural growth processes has always fascinated them.

One example of the mathematical exactitude with which plants use surfaces and space has already been given. In no way inferior is the astounding geometrical precision of plant forms, their 'superfinish', as the technician would call it.

33, 35   Of the three different 'thorns', in the photographs two are
34   man-made and one is natural. All three are so tiny that they can only be detected under the microscope, and all three are magnified about equally in the illustrations – approximately fifty times. The first shows the point of a very fine precision sewing needle of the highest grade. It does not taper off regularly, as would appear to the naked eye, but becomes sharper towards the top in a somewhat coarse and asymmetrical fashion, with the point curved out like a nose.

34   In contrast, the thorn in the second is technically perfect: it tapers to a point with the utmost regularity and is neatly rounded off. Here we have the topmost point of a fine cactus thorn several centimetres long. The precision of its form in the most microscopic detail challenges that of a record-player

sapphire needle. Technically the plant needle would be far 35
more difficult to produce because it is barely half as thick as the
crystal pick-up. Indeed, records have been played with cactus
thorns, as an experiment. The tone has been superb.

But how is this relevant to the theme of this book? It is the
basic material with which the plant solves its problems that
must be borne in mind. Consider what an unbelievable multiplicity of forms plants have evolved – large and small, some
roughly thrown together like the loops of lianas, others minute-
ly formed down to the most microscopic detail. And all out of
the one basic material, the living cell. Whereas we use iron and
steel, glass, cement, asbestos, paper, cardboard, polyester, glass
fibres, wire and rope, wood, sand and gravel, synthetic foam,
light metals, and the sapphire for precision microstructures,
the plant has nothing but cell material at its disposal. It solves
all its technical and ecological problems, often very similar to
our own, as well as or better than we do, with this material
alone.

The young cells of the most varied plant tissue are as like one
another as two peas. It is only during growth that they become
differentiated. They change, and change again, until they are
exactly adapted to the function they are to undertake. Optimal
collaboration is afforded by the single cells. Co-operation and
division of labour, in human thinking too often mutually exclusive, are so developed as to give the best possible adaptation
to the environment in which the plant has to live.

Perhaps it is precisely the absence of different kinds of material that compels the plant to use what it has rationally, to
pursue development till the best possible solution has been
reached in every case, and never to neglect the necessary feedback to the environment. And perhaps this too may be our own
best policy for the future: to cut out our extravagance as the
number of our raw materials diminishes.

CHAPTER SEVEN

Plant Transport

And God blessed them, and God said, Be fruitful and multiply, and replenish the earth, and subdue it.

Genesis, 1, 28

A good two-thirds of our planet's surface consists of water. 'Subdue the earth' also means, therefore, 'subdue the water'. Man has done it in the human way, the plant has done it in the plant's way. Man has succeeded in polluting and de-oxygenating brooks, rivers and even oceans on such a scale as to threaten life itself. Distinguished American scientists have calculated that by the year 1990, if filth and poison continue to pour into the seas of the world day and night, they will be biologically dead. By 1990 there will be no more fish in the sea. The algae, chief providers of atmospheric oxygen, will have died off, and terrestrial plants will not be able to supply the demand for oxygen. That is the human way of 'subduing' the water.

Plants, on the other hand, have learned how to inhabit the waters, how to purify them and to provide them with oxygen. In order to subdue the environment, man destroys it; with the same end in view as man, plants tend and maintain it.

Now it would be unreasonable to reproach man with not having made a living environment in the water. He is not a water-plant, nor a fish, and technological possibilities have their limits, although it is quite possible that terrestrial overcrowding may compel future generations to learn how to live and work under, or at least on, the water.

But these are Utopian considerations. Here I wish merely to compare the comparable. Let us consider the use of the water surface. Man has been able to move on water ever since the primitive dugouts of Stone Age times. But at what a cost! Today, insurance statistics show that between 300,000 and 500,000 gross tons of registered shipping are lost every year, counting only ships of 500 tons or more.

Plants, too, can navigate brooks, rivers, seas and oceans, but with infinitely more safety. Their boats are practically unsinkable; they survive collisions with drifting objects; and they withstand breakers and cliffs.

In the first place, plants are in possession of all the floating techniques known to man: the principle of the boat, i.e. of the hollow float open at the top; the principle of the pontoon, i.e. of the hollow, completely enclosed float; the principle of the raft,   36 which is not kept above water by large, interconnected hollow spaces, but by material which is itself lighter than water or which, like foam plastic, contains a mass of very small air spaces.

The leaves of the royal water-lily, described on p. 33, provide an outstanding example of the boat principle. True, they would also float as rafts (the young leaves of this plant actually make exclusive use of the raft principle), for they are themselves lighter than water. But their huge rim makes them excellent 'boats', whose ingenious construction ensures their capacity to carry quite heavy loads. As we have seen, one of these leaves can carry a girl weighing 90 lb without sinking or letting   6 in water. The great leaf, some 3 square yards in area, just floats about an inch lower in the water. In normal circumstances therefore a *Victoria amazonica* leaf can never drown.

The precautions taken by one of the best botanic marine floats strike one as similarly overdone, going well beyond the usual fivefold safety factor of human technology. This is the fruit of the coconut, ideally equipped for transport by ocean currents over bays and gulfs, from island to island, even across oceans, to win new terrain for this extraordinarily travel-minded shore-growing palm. The coconut floats in two ways, like a raft and like a pontoon. The familiar hard shell is encased in a thick sheath of tough fibre, elastic yet loose, and so light that it alone would keep the coconut afloat. On the outside, a smooth shell protects this sheath from damage. If this shell is destroyed in the surf or through rubbing against sand or rock, the coconut fibres (much in demand as carpeting material, because of their unusually hard-wearing qualities) protect the inner covering from abrasion. But should the fibres eventually be destroyed by a long sea voyage – coconuts often float for several months – the hard inner shell will always protect the central cavity, with its embryo, from being penetrated by water. The nut (which from the biological point of view is not a nut, but a stone-fruit) floats on as a pontoon. Thus the problem is solved of protecting the fleshy seed from every possible damage during its voyage towards distant shores, where it can germinate.

But this is by no means the sum of the measures taken for effective colonization overseas. When the coconut lands, it will certainly find no ready-made 'nest' in the shape of damp, fertile ground, for the chances are that it will be dumped by the breakers in some sandy, salty lagoon. But it carries its own food store – nourishing fruit pulp with rich fatty oils and much albumen, in sufficient quantity for the young seedling. Even a supply of fresh water, indispensable for germination and the first stage of growth, is not wanting: hence coconut milk.

Floating seeds and fruits like the coconut are to be found in plenty among shore plants. Although of fair size, they are all very light, as they have to be for dissemination by ocean currents. But on inland waters, as well, there is a great deal of plant navigation. Here water-lily seeds, especially, equipped with air-bladders, are carried by waves and currents to distant river-banks or lakesides.

On the whole, plants are passive when travelling by water. By availing themselves of the currents they can cover great distances without expenditure of energy. But there are active swimmers among them, moving quite purposefully through the water. The kind of propulsion that they use, far superior to any technical system hitherto in use on water-craft, is not unlike the stroke of a fish's tail-fins. Bacteria, the unicellular flagellates and the reproduction cells of many algae, fungi, mosses, ferns and other plants, move along actively in this way. Most of them are propelled by the highly complicated rowing movements of small flagella. Such propulsion is, like that of the fins of fish or the wings of birds, fascinating to any design engineer: the propulsive organ, whether it be flagellum, fin or wing, adapts its posture in every single phase of movement with the utmost precision to the prevailing currents. No technical system can even come near to doing this.

37

The blades of a ship or aircraft propeller, in spite of their variable pitch, are a first, and extremely poor, attempt at flexible propulsion. Professor Hertel, who has systematically taken nature as his model in aircraft design, has been experimenting for some years at Berlin Technological University with 'hover' propulsion in ships – a very crude imitation of the fin or flagellum stroke. He described the first results as surprisingly good. The efficiency of the still very rigid construction he rates at from 50 to 60 per cent. The mechanism behind the flagellum stroke, however, is incomparably better adapted to the move-

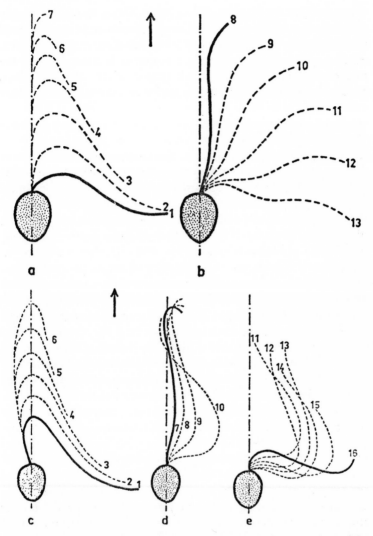

37 Adapting their double flagellum stroke exactly to the movement of the current, microscopic plant entities like the *Monas* (pictured here) achieve an efficiency level far superior to that of human technology. (a and c) Two different ways of retracting the flagellum; (b and d, e) two different types of forward stroke.

ments of currents, and probably attains values in the region of 100 per cent. It therefore uses, almost without waste, the propulsive energy at its disposal for the end in view – an unattainable dream even today in many areas of technology.

If necessity is the mother of invention, war is its father. It may be an exaggeration to say that only systematic preparation for destruction and genocide leads science and technology to great attainments, but the assertion is not entirely false. It was the desire to destroy that led to the invention of the catapult, which today peacefully launches the sporting glider; from warfare came rocket technique, which tomorrow perhaps will dominate the inter-continental transport system. It was military necessity that showed the way to the peaceful use of nuclear energy. True, the path of development pursued in this fashion is circuitous, expensive and dangerous. But to those who say that techniques which today are used peacefully would never have been developed without a previous military need, the answer is that this is because of man's inability to recognize the problems of everyday life, and to find constructive and flexible solutions. Plants, too, are experts in ballistics, although they make no war with bombs; plants, too, have developed slings, catapults, even airguns, and other explosive mechanisms, although they attack no one with them. They have achieved something which seems impossible for man: they have learned to shoot without needing war to father their invention.

Let us examine just three of the many truly ingenious shooting techniques of plants. One of the most interesting, because it is quite comparable to our airgun, is the method adopted by the seed capsules of many kinds of sphagnum moss for the dissemination of their spores. The capsules when almost ripe are twice as big as a pinhead, round as a bullet, and hollow. In the last stage of ripening they dry up and contract to about a quarter of their size. The original bullet shape disappears, and they become more like tiny gun-barrels, the tops of which are closed by a lid which can be blown off. Since the air trapped inside cannot escape, it is subjected to a pressure of 4 atmospheres as the capsule contracts – more than double the pressure in a motor-car tyre. Directly behind the lid, like a load of shot, lies a 'cartridge' of spores of the moss-plant. Towards the end of the shrinking process the lid suddenly bursts open, the compressed air is instantaneously freed, exactly as with an airgun, and the load of spores is fired. The report is heard quite clearly, and one can observe the recoil movement of the capsule. The tiny projectiles are shot up into the air, sometimes as high as 15 inches, and if the capsule is slightly angled, over 2 yards away. For a gun of only 2 mm bore, that is no mean achievement. Incident-

38 A range of over twelve yards and a 'muzzle velocity' of some 10 yards per second, are attained by the Mediterranean squirting cucumber (*Ecbellium elaterium*) when the ripened fruit separates from the stem and ejects the seeds.

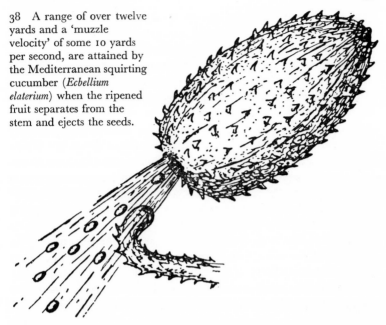

ally, it is not the range of the shot that matters, but its height, for the spores must be shot beyond the area of ground growth. The wind then looks after their dissemination. Therefore the little guns always have their muzzles pointed upwards, and shoot their charge vertically into the air.

The squirting cucumber of Mediterranean regions cannot 38 depend on the wind to carry its shot further, for it does not shoot spores as light as dust, but seed-grains bigger and heavier than the entire gun of the moss-plant. Therefore, it does not shoot vertically, but at the best average angle for a long shot – 50° to 55°. Physicists may disagree, claiming that the best angle for the longest possible shot is 45°, and this is mathematically correct, but the cucumber must take into account leaves that will be in the way of the little bullets if they fly too low. At an angle of over 50° the shot will tend to pass straight over such obstacles.

As the name indicates, the squirting cucumber works like a spray gun. The fruit resembles a longish plum in shape and size, and is enclosed in a firm, thorny skin. At maturity it detaches itself from the stem, and at the point where it breaks off an opening appears, through which a mixture of fruit juice and seed is almost instantaneously sprayed. This occurs because the contents of the berry are under about 6 atmospheres' pressure, and while it is being emptied its walls exert additional pressure.

A plant of the pumpkin family (*Cyclanthera explodens*) operates quite differently. Its fruit, which is about an inch wide, consists of two mussel-like shells clamped together and holding between them a curved, flexible lever. This lever grows from the plant at one end and, with its other end inside the two shells, loosely holds the seed to be shot out. Erectile tissue keeps the entire formation at great tension. Pressures of 14 to 16 atmospheres have been measured in these tissues, almost ten times that of a motor tyre. At the slightest touch or jolt the 'mussel' springs open, and the lever snaps out like a sling, shooting the seeds a distance of up to 3 yards.

Other plants work quite as successfully (for instance many varieties of *Dorstenia*) but at much lower pressures. They shoot their seeds out exactly as children shoot cherry-stones, squeezing them between two fingers till they spring out. The range of this compression artillery is from about 5 to 7 yards.

39   The dandelion's airborne seeds with their familiar parachutes are extremely light and specially constructed for transport by wind. When dryness, warmth and air currents combine to ensure that no mere brief gust is to be expected, but a fair and equable anabatic wind, only then do the flying seeds let go and venture on a trip. For this purpose the plant takes regular readings of relative air humidity, temperature and wind. This is why many trees scatter their pollen and their airborne seeds more particularly in the early afternoon, when there is much anabatic wind. The distance covered by the 'paratroops' is then greatest.

The similarity of dandelion seeds to parachutes is no accident. Such structures are able to be carried a long way by the wind and, on landing, the superstructure ensures that the seed, hanging under its umbrella, is deposited vertically, in the position most favourable to germination. Because of its elongated shape and the little barbed hooks at its upper end, the seed remains mostly in a vertical position in crevices on the ground, or in thick, low ground vegetation.

Parachute seeds like the dandelion's are to be found in various plants, and even in widely different botanical families. Here we have proof that a certain structure did not develop from the inner plan of a certain species or genus but that, irrespective of their family, the most diverse plants solved the same problem in the same way. Nor do seeds alone fly through the air on parachutes; so, frequently, do fruits. Thus two aeronautically similar structures must have originated anatomically in entirely different ways.

Parachuting is very far from being the whole of aviation. One can also rise in the air by balloon, aeroplane or helicopter. Man uses all these techniques. Is he then superior to the plant? No, for the plant uses the same techniques, and has even served as a model for human designers. Plants, moreover, have successfully experimented with some eccentric methods which are still beyond the bounds of aircraft technology.

Probably the most interesting example of plant aeronautics is the winged seed of a tropical variety of liana (*Zanonia macrocarpa*), which grows high up in the boughs of its supporting tree, and is notable for its garlands of beautiful, shining green foliage. Professor Haberlandt has described these strange fliers in almost lyrical terms:

Among them one sees the brown fruits hanging like great bells on high; if one waits till a gust of wind sets them in motion, suddenly it will seem as though a whole company of great gleaming satiny butterflies came whirring out of them. The big gourd-like fruit – its diameter is from 20 to 24 centimetres – springs open like a capsule on the downward-turned stalk, so that . . . a big triangular opening originates, at the edges of which the carpels strike inwards. So the opened fruit is like a great bell. The numerous winged seeds, stacked on top of one another to form a packet, are among the most beautiful and perfect things of their kind in existence. . . . They are easily torn at their delicate edges but their size and the lightness of the seed, which weighs scarcely a third of a gram, makes them highly efficient flying machines. Circling widely, and gracefully rocking to and fro, the seed sinks slowly, almost unwillingly, to the earth. It needs only a breath of wind to make it rival the butterflies in flight. (*op. cit.* pp. 140–1.)

Not only botanists have been impressed by the perfect flight of the zanonia glider. Five years after Professor Haberlandt's account (1893) the aviation pioneers, Ignatius Etrich and his son Igo, bought two aircraft – a glider and an ornithopter – which had belonged to Otto Lilienthal. From 1891 onwards, Lilienthal had made glides several hundred metres long with flying apparatus built by himself, and was the first to do so. In 1896, at the age of forty-eight, he had plunged to his death in the hills near Stölln. His successes were overshadowed by this tragic misfortune, but they could not be reversed. Two years after his death the Etrichs, father and son, textile manufacturers of Bohemia, decided to continue Lilienthal's work. Understandably, in view of the Lilienthal crash, they tried to make their models as safe as possible, but even so their very first glider crashed on its maiden flight in 1899. The two engineers refused to be discouraged, but they had learned from their mishap the necessity of researching aeronautically safe prototypes and studying them thoroughly. No such prototypes existed in technology. For years, therefore, Igo studied the anatomy and movements of flying animals, concentrating for a long while on bats because the structure of their wing membranes was easily copied. But their plans came to nothing because of the impossibility of copying the mobility of the bat's wing and the adaptability of its surface to the flow of air along it. So Igo Etrich looked for a rigid glider in nature, as a pattern for his models.

41, 42

Chance came to his aid. Professor Friedrich Ahlborn of Hamburg had just discovered the incomparable flying qualities displayed by the seed of *Zanonia macrocarpa*, and had demonstrated the importance of these qualities for the building of aircraft in a paper, published in 1897, on 'The Stability of Flying Machines'. This paper came to Igo's attention. He and his colleague Franz Wels took the next train to Hamburg, where he persuaded Ahlborn to give him a model of the seed and some valuable hints as to how the stability could be still further increased.

The Zanonia 'aircraft' is an all-wing glider, that is to say, a glider without a tail, and from 1904 to 1909 Etrich built all-wing gliders on this model. In 1906 Igo Etrich constructed a manned version of the model, and in 1909 he fitted it with a 40-h.p. engine. With the manned and motorized all-wing type the exact position of the centre of gravity was a source of difficulty; it had to be precisely adhered to in the interests of flight stability. In the model, the zanonia seed, this is always the case, but the very mobile human disposable load brought problems. For this reason the next Etrich model possessed a stabilizing tail, copied from the pigeon, and in May 1910 took off successfully on the first cross-country flight. It was an epoch-making achievement. The winged seed of a tropical liana had set the pattern for a crucial human experiment in air travel.

Plants, it will be seen, make use of parachutes as well as operating various types of glider. They also make use of the recoil principle in the launching of seeds and spores, just as it is applied in rocket technology, though in the case of plants this method is of secondary importance and small in scale. Otherwise, the principle would be uneconomical, for in fact plants tend to exploit the energy of the wind.

If this is to be done effectively, as large a working surface as possible must be offered to the force of the wind. This principle has been applied by man in the development of both the kite and the sailing-ship. It is also most effectively used by plants, whose so-called bladders combine the functions of balloon and box-kite. A good example of this is provided by the red balloons of the Chinese lantern (*Physalis alkekengi*). The skeleton of this plant forms an ideal lightweight construction for a box-kite, while its skin, which is taut and shining red in colour, is easily caught and tossed about by the wind, once it has been detached from the body of the plant.

Not all bladders are as big as the Chinese lantern. The seed of the long-headed poppy, though made not of one single air-bubble but of a number of very small cavities, is a superb box-kite of the tiniest proportions. With a diameter of only 0·7 mm, the little seed weighs no more than a tenth of a thousandth of a gram. The surface that it offers to the wind is made larger through a mesh-like structure in which pockets of air are caught, and this has the same effect as an increase in diameter.

In the case of larger flying bodies which would otherwise sink too quickly, nature has other devices. To keep the structure as light as possible, material is cut down. Thin skins, stretched over the finest of stiffening ribs, are in themselves very effective, but their efficiency is often increased by further methods. Like the helicopter, certain plants succeed in simulating, and presenting to the wind, working surfaces which for the most part do not exist. The seed of the Norway maple (*Acer platanoides*) has a surface of about 2 cm. Yet when dry it weighs only about 43 $\frac{1}{8}$ of a gram. The principle of the strutted 'wing' makes possible this advantageous proportion of surface to weight. When the winged seed falls from the tree, it begins to rotate quickly through air friction caused by its eccentric structure. The wing rotates in a spiral path approximately about its own centre of gravity, which is in the nut at its base. The motive principle of 44 this rotating movement is exactly the same as for the airfoil of an autogyro, and the effect is the same as for a helicopter landing in a glide when its fan blades will rotate of themselves from the air current. The rotation simulates a complete circular surface which the wind can grip and which, instead of the $\frac{1}{4}$ square inch of the maple seed, represents something over $2\frac{1}{2}$ square inches – an effective tenfold increase of surface attained by the simplest means. The rate of fall of such self-rotating seeds is slowed down by this artifice to an eighth or less of what it would otherwise be. Through this prolonged fall, a moderate side wind barely able to move the finest branches of trees will carry rotating seeds more than 100 yards from a tree 30 feet high. And eddying winds or up-winds, which are even more favourable to wide dissemination, will take them far further. Without this ingenious flight structure the seeds would fall more or less vertically from the tree, the young plants would have to germinate in the shade of the parent plant, and would compete with one another for light and space.

As one would expect in products of plant development, the various rotary wings are technically ideal in shape. Their rate of descent scarcely surpasses that judged to be optimal in a theoretically calculated wind-wheel, and is only 1·5 times as high as that of a hemispherical parachute of the same diameter, and therefore with a surface of over 6 square inches.

45 Another rotary wing seed, that of the ash tree (*Fraxinus
46 excelsior*), is not reinforced on one side like that of the maple, but is twisted on itself like the blade of an aircraft propeller. If the seed of the ash tree, proportionately magnified, is drawn
47 over a corresponding aircraft propeller type, it will be seen that the essential technical data completely coincide: the proportion of breadth to length is in both cases approximately 1 : 4·2, i.e. practically identical, and the angle of incidence is exactly the same at all corresponding points on both blades. In two respects, however, the ash seed does differ from the propeller blade: the broadest part of the blade surface goes much further out (as seen from the centre of rotation) than in the aircraft propeller, and the blade of the ash seed from the first third outwards is proportionately much thinner than the propeller blade. Both these features are definitely to the plant's advantage. Greater breadth of blade in the regions of greater rotating speed means a greater working surface for the airstream; and the greater general thinness means a considerable economy in

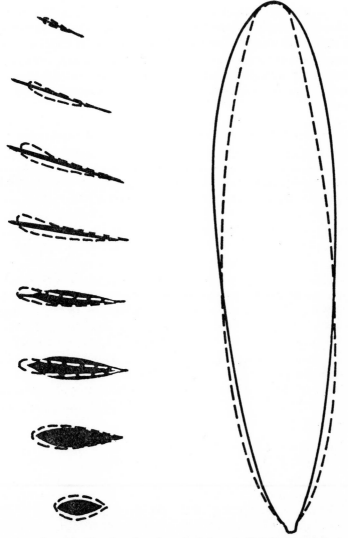

47 Aerodynamically, the winged seed of the ash (full line) and the propeller blade of a large commercial aircraft (broken line), superimposed and
drawn to the same length, are in remarkable agreement. Left, the sections of
both shown at varying levels.

weight – always an important consideration in the construction
of missiles. Why then do our aircraft technicians not build to
such advantage? Because it would mean a loss of stability.

Two more types of plant 'aircraft', may be mentioned here,

for the sake of completeness. The first is a delicate disk shape rather like a flying saucer, with the seed or fruit in the middle of the disk; the second is *Dryobalanops*, the winged seed of which resembles a shuttlecock. This is not a particularly impressive flyer, being intended primarily as a parachute, so that the seed does not hit the ground hard, but settles gently and without harm.

As this is not a specialist botanical textbook, a brief digression on a subject of interest to amateur photographers may be in order at this point. If you enjoy tricky problems for the camera, try to photograph a self-rotating flying seed in descent. Here are a few tips. Be sure to use a tripod, and a lens with as wide an angle as possible. As backdrop, any dull black surface is suitable, for example fine black woollen stuff. Set up the camera in such a way that the sharply defined zone (with the stop fully open) is about 8 inches in front of the black background, and the image field a good postcard size. Such focusing is best for judging later whether the seed has fallen past at approximately the right point. The field should be brightly and evenly illuminated on both sides. For this two strong diascopes are useful. Light must not fall on the background. To make the falling seed stand out against the background, colour it evenly with Chinese ink (as thinly as can be, so as not to change the weight). Finally, mark on the ceiling of the room the point vertically above the middle of the image plane; this is the point from which the seed should be dropped.

After these relatively simple preparations you need only have patience and a great deal – a very great deal – of film. In my experience the result goes something like this: in more than 60 per cent of the falls the seed does not fly past the lens, but goes its own way. In most of the rest the flight path just grazes the edge of the image field, or is quite outside the plane of sharp definition; or the shortness of the fall from the ceiling does not allow a beautiful regular rotation to develop. Or grains of dust are also conveyed on the air, leaving ugly white streaks on the film. About every hundredth shot is a good one. Which is what makes the photographic gamble so fascinating.

Transport by wind automatically suggests balloons, autogyros, parachutes, and all kinds of flying bodies. But wind transport need not always be airborne: sand yachting, for example, uses

1   A poplar leaf which has been masked in the middle by a round disk for twenty-four hours, and then 'developed' with a solution of iodine, shows clearly where the leaf has been affected by sunlight, generating starch (the dark area), and where it has not (the circular pale area).

2　Following the sunlight has caused the foliage on an ivy-covered wall to grow in such a way as to form a thick mosaic, the leaves covering and shadowing one another as little as possible.

3   The 'window plant' *Fenestraria* grows almost completely underground in South Africa, its natural habitat. Only the transparent 'windows' at the tips of the leaves admit the sunlight. This specimen comes from a German garden, and because of the different light and temperature conditions of Europe it has a slightly altered appearance.

4   The inner surface is so transparent that typewritten letters can clearly be read through the leaf. Bright sunlight streaming through the 'window' (right), into the leaf, is refracted and scattered in the clear watery interior, and reaches the chlorophyll-containing walls.

5　In nature rubbish tips like this do not exist. Human garbage disfigures landscape and endangers health. But plants recycle their waste with an economy we are only beginning to appreciate.

6 The great floating leaves of the royal water-lily (*Victoria amazonica*) with their vertical rims are an example of botanical 'boats'. Their load-bearing capacity is surprisingly high: the girl in the photograph weighs nearly 90 lb.

7 An ingenious form of rib construction on the underside ensures the stability of the great *Victoria amazonica* leaf.

8 The architect and former gardener, Sir Joseph Paxton, built the Crystal Palace exactly on the model of the tropical water-lily leaf.

9, 10 Corrugation of a thin sheet of paper multiplies the carrying capacity more than a hundred times. The giant leaves of many fan palms owe their stability to the same principle. The leaves of the Chinese fan palm, *Livistona chinensis*, shown here, do not lose any stability if they go in at the edge at every second fold. This partial cutting away of the leaf, which in no way detracts from its biological functions, is a natural feature, and in many species the leaf will automatically lose some of its substance like this as it grows.

11   Like an enormous palm leaf, the pleated cantilever roofing of the entrance to the Mont Blanc tunnel is 'torn' at every second fold.

12   The iron reinforcement cage (far left) of a ferro-concrete column. The centre of the cage is hollow, and the reinforcing structure will underlie the finished surface of the column.

13   The solid reinforcement material of a dead saguaro cactus (left) has exactly the same arrangement as the cage of the concrete column: the material lies immediately under the outer face of the living plant column, and there is none at the centre.

14   The trellis-like framework of the prickly pear, *Opuntia*, is similarly hollow inside.

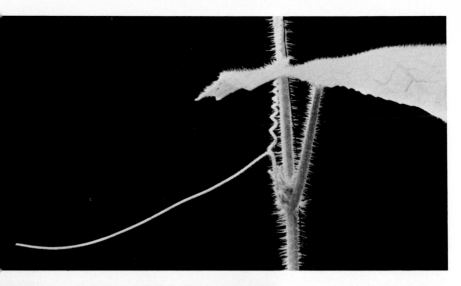

15 The 6-inch tendril of the wax gourd carries out slow, circular search movements to find a hold.

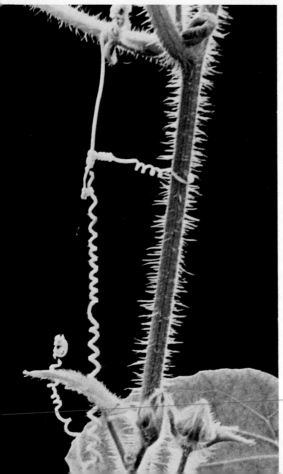

16 Once the tendrils of the wax gourd have found a hold, as on this wire, they coil themselves exactly like a spring. This makes a firm and yet elastic connection.

17 Reconstruction of a Bronze Age lakeside village built on piles. The piles ensure good aeration and protect the huts from decay and floods. At the same time they are an ideal anchorage in marshy ground.

18 Like the mangrove, the screw-pine, *Pandanus utilis*, is another of nature's pile-dwellings.

19 Gravel pits yield building material at terrible cost in countryside dereliction and blight. Only a year before this photograph was taken, the people living in this little German town could enjoy the woods beside their homes. Nature provides the model for an alternative with skeleton construction.

20 Interlacing, mutually supporting branches give the same stability to tropical fig-trees (*Ficus rumphii*) ...

21 ... as to modern steel frame buildings or ...

22... the filigree-like fruits of the Chinese lantern (the thin, shiny red skin of the bladder has been etched away).

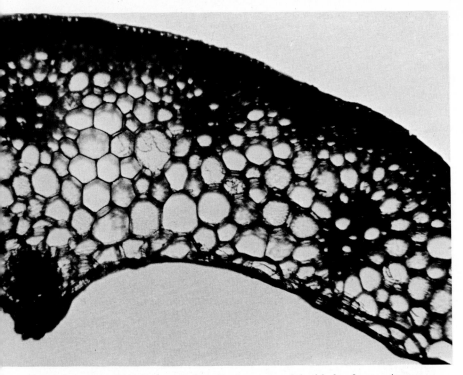

23 The microscope unveils the secret of the blade of grass. A cross-section of a barley stalk, multiplied 70 times, shows how the stem wall, only about 0.6 mm thick, owes its considerable stability to the technologically modern device of a honeycomb sandwich.

24 Aircraft construction achieves lightness of weight and great stability in the same way as the blade of grass: by honeycomb sandwich construction.

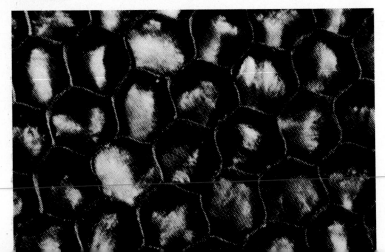

the force of the wind for propulsion. Where there are wide stretches of sand and enough wind, this has proved to be a popular sport. And plants in a similar environment practise the same kind of locomotion. For them, however, to be driven by the wind over the dunes is no game, but a means of gaining new living space.

In India, when the vegetation of the dunes on the seashore is withered by the powerful easterly monsoon wind, and almost every plant is dry and leafless from the long drought, the stiff, bluish dune grass (*Spinifex squarrosus*) sends its 'offspring' in search of territory. In the course of the year this grass develops remarkable fruit as big as a human head, feather-light elastic balls of a perfectly spherical shape, containing in their centres a fair number of tightly-packed ears. When the monsoon wind sweeps over the sand it catches these balls and chases them helter-skelter over the dunes. This kind of 'sand yachting' guarantees the plant ideal dissemination. The seeds are shaken out of the balls as they roll, and frequently distribute themselves over large areas.

The same method of dissemination has been developed by plants on the steppes of southern Russia and Asia, where the natives call these shapes scurrying before the wind 'steppe witches'.

In early May 1934 Helmut Rempe, a young biologist working on his Doctor's thesis, arrived on the tiny island of Heligoland in the North Sea to look for the pollen of firs, pines, oaks and birches. This seems at first sight an odd undertaking, for none of these trees were present on Heligoland in 1934, and indeed the only ones there at all in May of that year were a single elm and a few willows. The nearest point on the mainland was 32 miles away, and the nearest island (Scharhörn) 27 miles. But Helmut Rempe knew what he was doing. At the northwest tip of the island, close to the cliff edge 160 feet above sea level, he hung small, vaseline-smeared rolls of paper 14 mm in diameter and 45 mm long at the top of a six-foot pole, and waited. Every twelve hours he changed the rolls, making a total of seven times in all. The result was surprising. After the period of three and a half days was over, the student counted 955 grains of oak pollen for every square cm of trap surface – almost ten grains per square mm. This would be quite enough to pollinate an oak tree of female blooms on Heligoland.

Where did the pollen come from? The only possible answer is that it was blown at least 40 miles over the sea from the mainland.

Pollen of the other species (pine, fir and birch) reached Heligoland in somewhat smaller quantities in the course of the experiment, but these also would have sufficed to pollinate trees growing there. Rempe sums up as follows in his thesis:

The following figures will demonstrate that the quantities of pollen blown at least 40 miles across the sea from the mainland over the brief duration of the experiment were extremely large. Over the full period of 84 hours, a total of some 27 million pollen grains would have passed through a frame one yard square on Heligoland. The twelve daylight hours of 4 May 1934 would alone account for 15 million pollen grains.

Plants, therefore, can carry out extensive cross-country flights. A range of even several hundred miles is not uncommon. A grain of birch pollen, for instance, carried up by warm air currents to a height of 6,000 feet, will sink so slowly by reason of its lightness and comparatively large surface area that a Force Three wind (at 11 m.p.h. just strong enough to move leaves but not twigs) would carry it 250 miles before it reached the ground. For such a long flight the pollen grain would need several days, and one cannot count on a steady wind for all that time. But, on the other hand, neither up-winds nor violent storms have entered into our calculations, and these are responsible for the long-distance transport of a considerable quantity of pollen.

Moreover, in good flying weather clouds of pollen at heights of 18,000 feet or more – almost the flying height of modern jets – are not uncommon. The tiny particles are so light that even in absolutely still air they would take 66 hours (in the case of the birch pollen) to fall to the ground from such a height. On the other hand the maple seed, which works on the autogyro principle as described in a previous section, would get down in about two hours from 18,000 feet. Compared with this, a parachutist descending from a similar height would seem to be plunging down: he would reach the ground in 20 minutes.

The capacity of pollen to float in the air is therefore quite extraordinary. This is why the particles are so microscopically small. The pollen grain of a fir tree has a diameter of about 0·15 mm, while that of a horse-chestnut and of most varieties

of willow are only a tenth of that. The seeds of many varieties
of orchid, which are likewise carried by the wind, are incredibly
tiny: half a million together weigh only one gram. But they are
gigantic in comparison with fungus spores, whose diameter
measures only about 5/1000 mm. Twenty billion of them are
needed to make up one gram. Naturally they sink through the
air far more slowly even than pollen. In a free fall from 18,000
feet in absolutely still air they would take not less than 370
hours, or exactly half a month, to reach the ground.

This pollen whirling through the air, these hovering seeds
and far-travelling fungus spores, are often perceived as nothing
more than wind-borne dust. Yet what is it that is being so
effortlessly transported over immense distances in each grain?
nothing less than the detailed blueprint, together with the exact
growth and behaviour programme, for a whole plant, fungus,
orchid or tree. Thus billions of information stores, each one
more efficient than an electronic data bank of the most advanced construction, are being invisibly conveyed through the
air, often at great heights. In Chapter 10 I shall enter into
greater detail on a performance that no technology can hope
to emulate.

These data banks are so small, and at the same time are produced in such unimaginable quantities, that they can exploit the
subtle transport conditions constituted by dust-in-the-wind
flight and make use of so cheap and advantageous a long-
distance transport system. Incidentally, as we have seen, the
plant does not squander material haphazardly through excessive production: flowering trees do not give off their pollen
at a constant level throughout the day, but prefer the early
afternoon hours in which they can rely on a favourable up-wind.
It has already been shown how, in solving a problem, plants
will exploit every possibility. In the context of dissemination
methods this means that, as well as travelling by water and air,
plants also make use of flying and running animals. What could
be more convenient for the plant than to join an animal which
is constantly on the move in its flight or its wanderings?

According to the type of transportation used, the plant may
travel as a 'paying passenger' or as a 'stowaway'. Aircraft are
not easy to board unobserved: if you use them, you must pay
for the journey. On the other hand, you can occasionally travel
economically on land vehicles, having jumped on unobserved.
It is much the same with plants: they pay for dissemination

by birds, but usually travel as 'stowaways' with land animals by clinging to their coats. The sweet, tasty pulp common to almost all berries and stone-fruits attracts birds, and is their 'pay' for distributing the seed. They eat the juicy fruit, but cannot digest the hard stone inside, which they deposit somewhere far away from the parent plant. In this way many plants have bridged the gaps between islands in the dissemination of their species.

For many epiphytes – plants that live on other plants, e.g. in the branches of trees, without sponging on these plants for food or otherwise harming them – seed distribution by the wind and, above all, by birds is essential for survival. How else would the seeds get into the high branches of the supporting plant?

The most ingenious way of securing the services of birds has been developed by a group of tropical plants. Professor Haberlandt has described their manoeuvre in the following words:

> In the tropics, even more than in temperate regions, birds and other, larger animals are recruited for the distribution of seed. Very little is known as yet of the diverse adaptations to this end. I will merely draw attention here to the seeds of some leguminous plants, which obviously imitate the appetizing colours of berries, in order to attract birds. The best-known example is the rosary pea (*Abrus precatorius*). Still more striking are the shiny crimson seeds of *Adenanthera pavonina*, standing out boldly against the pod, which rolls up and turns its bright yellow inner side outwards. The most magnificent, however, are the great seed-pods of *Pahudia javanica*, about 4 inches long and 2½ inches wide, when the enormous jet-black seeds with their scarlet casings are viewed against the gleaming silvery inner skin of the open pod. A more effective colour contrast could scarcely be imagined. That this tactic is designed to attract birds and deceive them into swallowing the indigestible seeds, or at least strewing them around, seems all the more probable when one considers that none of these seeds fall to the ground immediately the pods spring open, but continue to adhere to the open flaps. (*Op. cit.*, p. 142.)

Seeds, fruits, and even whole parts of plants frequently travel on land animals as 'stowaways' without any deceptive manoeuvre. They simply cling to their coats as they pass, and are brushed off again by the undergrowth, often many miles farther on. This kind of travel demands the development of special gripping devices. The grip must be quick, tight, and secure, but at the same time quick and easy to detach.

It is only a few years ago that human technology discovered a

system of this sort in the Velcro fastener. As a reliable yet easy- 48
to-open fastening it quickly became popular in the form of
'touch and close' tape. The pillow-covers which have to be
changed frequently in modern passenger aircraft are fastened
with it, department stores particularly recommend it for the
easy hanging of curtains, the clothing industry often substitutes
it for the well-tried zip, and the photographic trade uses it in
making adjustable cases for camera accessories. The catalogue
of uses for the universal, simple Velcro fastening is almost infinite. I do not know whether or not the inventor of this
principle, as novel for man as it is simple, worked consciously
from an age-old model in nature. The German manufacturer
at least, in the names *Klettverschluss* ('bur-fastener') and *Klettband*
('bur-tape'), made unequivocal reference to the plant which has
always practised this, the burdock.   49

The whole fruit of the burdock is specially constructed for the
particular task of dissemination by furry animals. When its
barbed hooks fasten on a passing animal it detaches itself as a
whole quite easily and without damage. But when the bur is
brushed off the coat of the carrier, the casing splits open. Quite
often, pieces fall off at intervals, so that the burdock seeds are
released from their container over the widest possible area,
and do not later germinate all in the same place.

'Stowaways' employing barbs and gripping mechanisms can
be counted in their hundreds in the plant kingdom. An
interesting example is the 'harpoon' of the various species of 50
bur marigold (genus *Bidens*). Though their barbs are only a
few millimetres in length, nobody who has made their acquain-
tance will easily forget these obstinate seeds. It takes a strong
brush and a lot of energy to get them out of your clothes. Under
magnification, the top half of one such barb closely resembles 51
the bone harpoon heads carved by our paleolithic ancestors for 52a, 52b
hunting animals.

In the development of efficient structures, whether Velcro
tape or harpoon, man is hard put to it to discover any techniques not already used by plants. But how many techniques
does the plant have at its command that man does not yet know
or use? Thousands of years went by before we took over the
skills of the burdock; until recently we were oblivious of its
significance, though children see and play with it each autumn.
How many simple and cheap solutions to everyday problems
may still be lying unrecognized before our very eyes?

Seeds and fruits are not alone in exploiting the transport possibilities of animals: sometimes whole plants take to the road with them. Desert and steppe plants in particular, which have learnt to survive for long periods without moisture, make great use of this method. They survive journeys of days or weeks in the coat of their carriers because they can do without water. In America there are varieties of the *Epiphyllum* cactus whose individual shoots bear crowds of devilish barbed hooks that claw at almost anything that passes. Even hard, highly-polished leather is not too slippery for them. Especially during long droughts, single stems or whole groups of stems detach themselves very easily from the plant and let themselves be carried off. They may be taken somewhere moister where they will have a chance of survival. When they are brushed off or simply fall off through jolting, these stems quickly take root and continue their growth. The 'jumping cholla', too, a cactus of Mexico, is particularly good at travelling as a 'stowaway'. It actually 'jumps' onto passing animals that have not quite touched it, possibly through electrostatic attraction.

Whether in the coconut crossing the sea, or the grass-balls of India racing over the dunes, or the 'steppe witches' of southern Russia disseminating themselves in similar fashion, there seems to be a marked endeavour on the part of plants to gain territory for the propagation of the species. That holds good even for the long-distance transport of pollen in the wind. In this last case, propagation of the species might at first seem out of the question, since pollination presupposes that a specimen of the same species must already be growing and flowering at the other end. And yet, through cross-pollination followed by re-crossing with other, related species, a real propagation of the species can take place even here.

What do plants gain by their wanderlust? Does it matter where they grow? It might be argued that the place where the parent plant blooms and flourishes would provide the best conditions for its offspring, while the further the seed, fruit or part of the plant travels from the parent, the more difficult it becomes to guarantee a favourable environment. As a general principle, that is true. But there are two very important reasons for the advance into new territories. First, by emigration plant species create colonies where one day they may be able to survive after life in the native territory has become impossible.

Thus, for example, many alpine plant species only survived the
great ice ages because single plants had migrated in time to
warmer valleys or into the plains. Nor does there need to be
an ice age for catastrophe to threaten existence. A gradual rise
of the average temperature by only $\frac{1}{2}$°C, together with a slight
increase in precipitation, can limit the chances of survival for
many plant species. The mighty saguaro cacti of Arizona take
in too much water as a result of such a moderate change of
climate. They suck themselves full, then burst and die.

There is a second, and perhaps even more important reason
for as vigorous a distribution drive as possible: the avoidance of
monocultures. If all seeds were to germinate in the vicinity of
the parent plant, there would simply be more regions in the
world with a completely uniform flora. There would be no
possibility of a mixed vegetation growth, which alone leads to
natural communities and is so important for the continued
existence of all. Monocultures would grow up, and these in the
end are not viable. Epidemics would transform great stretches
of country into barren wastes in a very short time, and wind
and water would carry away the humus layer, thus making new growth difficult. Healthy undergrowth would not
develop in forest monocultures. In periods of drought there
would not be a sufficient reserve of moisture: whole forests
would wither away. Storms would flatten whole tracts of woodland, and trees felled by the wind would be a welcome feast for
the bark-beetle, which would then multiply rapidly and destroy
also those neighbouring tree populations not affected by the
storm. Nature effectively avoids all these devastating phenomena through mixed growth. The advantages of botanic communities are therefore obvious. But man, in his ventures into
forestry, has not only failed to note these advantages; he has
positively ignored them.

Any thorough history of the island of Madeira is likely to include an indictment of its first settlers, who burnt down the
thick and extensive woods so that now barely a trace remains;
and in descriptions of the northern Mediterranean lands there
are frequently references to the irresponsible Italians and
Yugoslavs who, centuries ago, denuded their highlands of trees,
and today, however much they may try, cannot succeed in re-
afforesting them.

But instead of pontificating on the past, there is today a clear
need to send out warnings of the havoc being systematically

prepared for the future in all countries of the temperate zones. For our forestry is cultivating on the grandest scale what nature seeks by every means to avoid: monocultures. The dangers characteristic of such cultures – wind damage, disease, destruction of whole forests by drought – are no longer a mere threat: they have long become a reality. On 13 November 1972 the wind flattened vast expanses of monoculture forest in West Germany within a few hours: over an area of 250,000 acres alone, 40 million trees lay uprooted or broken off on the ground. This worked out at about 18 million cubic yards of matchwood.

In order to clear the enormous quantities of fallen timber as quickly as possible, the State summoned lumberjacks and foresters from the Tyrol and even from Canada to its aid. The great risk was in delay: the dying wood might well lead to a bark-beetle epidemic, dangerous to the trees still standing. In spite of rapid and intensive large-scale operations, commercial timber that had been cultivated for decades was turned into firewood by an hour's storm.

Was this an unforeseeable, unavoidable catastrophe? In an area bordering on that of the 1972 damage a storm in 1958 had cut a broad swathe in what had been an old stock of firs. What measures did those responsible take then? The local authorities made a nature trail, took it past the unwanted clearing and put up a notice explaining how wind damage was to be avoided by planting a mixture of shallow- and deep-rooting trees. Shallow-rooting species like the fir, it was explained, were at risk from storms; the pine, the spruce, the oak, the lime and the European larch were more resistant. But today a newly planted stock of trees may be seen growing on the recently damaged area – a new fir monoculture! And behind that, as the finishing touch to this sad picture, the dumb and broken witnesses of the 1972 storm. Will yet another monoculture be cultivated in this desolation, to fall a certain victim to a storm in twenty, thirty or forty years? In view of the discrepancy between the intelligent message of the nature trail notice-board and the pure fir stock behind it, one can scarcely hope for anything else: 'They know not what they do' – and they practise not what they preach.

Monocultures are unhealthy. Nature therefore prevents them by every possible means. But we human beings are never tired of promoting them. Can we wonder then that we must so often lament the consequences of our imprudence?

CHAPTER EIGHT

Masters of Hydraulics

The task of carrying 40 gallons of water up almost 60 feet – to an apartment on the sixth floor, for instance – is an undeniably strenuous undertaking. And yet it is no more than a full-grown birch does on any warm summer's day. A beech coppice only 100 yards square contains around 400 trees 75 to 90 feet high. The leaves of this little wood evaporate an average of 20 tons of water in a single day – equivalent to the capacity of a large tanker truck. Before this water evaporates it is lifted an average of 60 feet through the trunks, branches and twigs. Anyone who thinks this a trifle would do well to work out how many buckets it would mean, to which floor. Quite a considerable achievement. The surprising thing is that the plants themselves need supply no energy to do this. The appropriate structural measures mean that everything happens automatically. The evaporation from the surface of the leaves causes a constant compensatory suction of water. This suction communicates itself through twigs, branches and trunk down into the roots. The force of the suction caused by evaporation from the stomata of the leaves is tremendously strong. When, on a dry summer's day, the surrounding atmosphere has a relative humidity of 45 per cent, the evaporation suction corresponds to the pull exerted on a rope 3 mm in diameter by an eleven-stone man suspended from it: 1,000 atmospheres! Thus evaporation literally draws the water from the leaves by force, and pulls more water up through the plants to replace it. The driving force is, in the final instance, the sun. Plants make direct use of its energy for water haulage. Once again we see how simply, and at the same time directly, plants use the almost unlimited energy that is available.

But are they not in fact wasting energy here? The frictional resistance in these finest of all water conduits is so great that only 2 per cent or less of the 1,000 atmospheres of suctional pull operating at the stomata can still be measured in the roots. But it is precisely in this apparent wastage that the plant's real technical feat lies, for only by dividing up the 'water-pipes' into a great number of microscopic tubes, a few thousandths of a millimetre or less in diameter, does it accomplish what no man-made suction pump has ever managed: the upward suction of water for more than thirty feet. Columns of water pumped higher than this in ordinary man-made pipes inevitably collapse from physical causes. Without the development of capillaries, the world's finest 'water-pipes', no tree could grow higher than thirty feet. They alone make possible the huge growths of over 300 feet. The highest tree at present in America, a Californian giant sequoia, measured 362 feet in 1967. Australian eucalyptus species grow even taller. Dr Ferdinand Müller is reported to have found one specimen 510 feet high, by the Latrobe river. The lowest boughs of this giant were about 300 feet above the ground. When Australian lumberjacks, in 1950 or thereabouts, felled a mighty kauri pine in the uplands of Queensland, the bare trunk, after the crown and all the boughs had been removed, was still over 270 feet long. It had a circumference of some 20 feet and weighed 24 tons.

The botanist Professor Mohr has calculated that, to get water to the top of a 450-foot tree, taking into account the force of gravity and the frictional resistance in the fine sap ducts, the single threads of water in the trunk are exposed to a pull of 35 atmospheres. The resulting suction in the capillaries is so strong that they contract. As innumerable sap ducts run side by side in a tree-trunk, one might imagine that the trunk as a whole would become noticeably thinner in the morning, when the rate of evaporation increases greatly against that of the preceding night. Indeed, daily variations have been observed in the diameter of the trunks of Californian giant sequoias (*Sequoiadendrum giganteum*) and Monterey pines (*Pinus radiata*). The evaporation is greatest, and the tree-trunk thinnest, towards 2 p.m.; around 4 a.m. the opposite is true.

The fineness of the little ducts naturally means that the sap can rise only relatively slowly in the trunks. The rate has been measured as 3–6 feet an hour in coniferous trees. It is interesting that plants of low stature, which can indulge in the luxury of thicker water-pipes, promptly do so. Accordingly, the haulage speed of their sap is greater. A stream of between 120 and 150 feet an hour was ascertained in leaves of wheat.

One last figure, to conclude these brief mathematical reflections on the theme of water transport: in forest regions, through evaporation from the leaves, 60 to 70 per cent of the annual rainfall is returned to the atmosphere. Extensive forest stands thus make an invaluable contribution to climatic equilibrium. Can we really afford to go on recklessly cutting down hundreds of thousands of square miles of forest throughout the world each year, and building in its place streets, housing estates and factories, which in many cases could have been built elsewhere? Can we continue, year after year, to destroy tens of thousands of square miles of woodland, merely to exploit the thin layer of sand and gravel beneath it? In view of the forest's function as a climatic factor, its still more important job of purifying the air and enriching it with oxygen, and, not least, its recuperative value for millions of tired town-dwellers, the answer should not be hard to find.

On an average, 80 per cent of a plant is water. In typical dry-region plants the proportion of water is lower, and in water-storing plants it can amount to 95 per cent. Water is of the highest importance for the plant, as for all living creatures. It regulates the firmness of the tissue; it serves as a solvent for the nutrient salts that have to be conveyed in the plant; it has a direct influence on electrical processes in the plant; it is important for all chemical reactions in the organism; and it plays a decisive part in the building up of firm, non-watery plant substances. The water supply is accordingly one of the central problems of the plant's existence. Aquatic plants are favourably placed in this respect. They can absorb the all-important moisture through their entire surface. Terrestrial plants get their water as a rule from the damp earth through root hairs. The structure of these fine hairs, which are particularly active, is eminently practical, and can even undergo adaptation in one and the same plant. If, for example, a plant cultivated in the soil is transplanted into a completely earth-free, watery nutrient solution, the form of the roots changes in a very short time. The fine little ramifications of the roots are admirably adapted to their task of actively absorbing moisture and passing it on into the ducts of the plant.

How efficient a system really is can best be judged in border situations, at the point of transition from the still barely possible to the impossible. Necessity is the mother of invention. It should not, therefore, be surprising that plants' most interesting and elaborate techniques are to be found where scarcely any water is available.

One type of environment in which water is definitely a scarce commodity is in areas where plants come into direct contact with neither earth nor water. Most tropical orchids, for example, flourish under these conditions, growing high up in the branches of jungle trees, to which they simply cling without drawing moisture or nutriment from them. There are no roots reaching down to the damp earth, and the trunks and leaves of the supporting trees are often so smooth that all rainwater runs off them immediately. Running water therefore is hardly available at all, certainly not in sufficient quantity. But the air in the orchids' native forests is moist. Frequent rainfall, and heavy evaporation through the leaves of the trees in these high temperatures, create a regular hothouse atmosphere. Therefore the moisture essential to life is literally in the air, and the orchids appropriate it in the ordinary way: they develop roots, which hang freely in space, ramify, and often form complete networks. But the roots lack the enveloping damp earth that prevents them from drying up, and enables them to absorb the moisture. The plants' solution is simple and effective: they make an artificial 'earth' for themselves. The aerial roots of orchids are embedded in a thick, loose tissue, known as velamen, which is porous and consists of dead cells. When dry it can easily be compressed, and looks quite white because of the many air-filled cavities. Velamen sucks up the slightest damp in the air as greedily as blotting paper. When very damp, it becomes quite dark in colour. The roots can easily absorb water directly from this sponge and channel it to the plant's water supply system.

Quite different in structure, but very similar in function, is the action of many tropical milkweeds, which likewise grow in the branches of jungle trees. They also protect their roots from drying up by ensuring that there is vapour-saturated air in their immediate vicinity. But instead of forming a special spongy tissue, these plants cover the roots, which lie close to the supporting tree, with the shade of their leaves. The closer the leaves lie to the bark, the damper, naturally, is the air between. A representative of this plant family commonly found in Java (*Conchophyllum imbricatum*) shows this protective arrangement in a highly developed form. The rather thick, fleshy leaves press against the bark of the host tree, down which grows the thin stem of the plant. Thus under every leaf there is a damp hollow, in which a root sprung from the stem grows, and ramifies abundantly. The coinciding position of the leaves and root clusters is therefore no accident.

The pitcher-plant (*Dischidia rafflesiana*), another species of liana of the same family, and also a native of Java, has further developed the principle of the damp root space. Its stems produce two kinds of foliage, on the one hand quite normal green leaves, on the other pale yellow-green structures modified to form strange sacs which look from the outside like longish, flat tubers. The inner wall of these is lined with a thick, purplish-black coat of wax, with tiny slits through which the pitcher-plant's leaves can evaporate moisture. At exactly the points where the leaves appear, roots also sprout. Here is the explanation of these peculiar leaf sacs, unique in the vegetable kingdom. The entire structure is, in four separate ways, an outstanding apparatus for the utilization and conservation of water resources. First, as a result of temperature variations between night and day, shadow and sunlight, water is easily condensed on the inner walls of the bags and directly absorbed by the roots lying against them. This principle of condensation through the alternation of heat and cold is exploited by farmers on many dry volcanic islands, for example Lanzarote in the Canaries: they cover their fields with a layer some 8 inches thick of porous pumice gravel or lava ash, which the islanders call *jable* or *chahorra*. The cool night temperatures cause moisture to condense in the cavities of this material, from which the growing plants profit. Without such a protective layer of porous material agriculture would be impossible in these regions, where sometimes there is no rainfall at all for up to three years.

Secondly, in the interior of the pitcher-plant leaf the relative degree of humidity even during a drought is still so high that the roots do not wither. Furthermore, the total quantity of water given off by evaporation from the tubular leaves is reduced to a minimum, for the air inside these bags is always damp and still, two conditions that considerably limit transpiration. Finally, moisture given off condenses immediately and can once more be absorbed by the roots. One is vividly reminded of the water-supply systems of many industrial nations, where the water of certain rivers is treated again and again, piped back to households, and thus drunk as much as five times over.

But where man treats water by adding chemicals to it,

Dischidia uses a procedure physiologically far less objection-
able: distillation. There is no better way of recovering used
water. In the language of the environmentalists, a complete
recycling occurs in the pitcher-leaves of a tropical liana.

What the pitcher-leaves can do, we too must learn to do, for
we are in the process of poisoning all the water of the earth.
Under the title *What the future will bring*, the Weltbund zum
Schutze des Lebens (World League for the Defence of Life),
an international organization with its headquarters in Luxem-
bourg, lately published some alarming forecasts:

A group of US scientists fed data on the present to a computer, then
put questions about the future to it.

In 1977 there will be no river the water of which can be used for
any purpose whatsoever.

In 1980 there will be no more natural drinking water.

In 1990 the seas will be biologically dead through pollution. Yet
the future depends on oxygen production by the seas, because the
continental covering of green plants no longer suffices to meet the
high consumption of oxygen. No life without breathing! The seas of
the world are a field that needs no cultivation, yet would be capable
of providing the whole of the human race with valuable albumen.
Meanwhile filth and poison run day and night from all the rivers of
the earth into the seas, and the responsible authorities appear to be
incapable of doing anything but acknowledging this fact.

In 1995 most of the foodstuffs entrusted to us by nature will have
become uneatable or unprocurable. It will no longer be possible to
produce sufficient synthetic drinking water. If for a brief space there
is sufficient, it will only be because the number of human beings has
already considerably diminished.

In the year 2000 life for human beings will no longer be worth
living. The world of animals and vegetation will come to an end. In
landscapes traversed by irreclaimably contaminated rivers, on
ground made infertile by artificial manures and countless poisons,
in an atmosphere which has long lacked sufficient oxygen but contains
in its place unimaginable quantities of the most noxious gases, no
tree, no shrub, no flower will be able to thrive, nor any animal exist.

In the year 2010 the end of the process started by human arro-
gance, short-sightedness, greed and morbid lust for power will be
clearly in sight. The earth's atmosphere will have become un-
breathable for even the few things left alive.

To sum up: the point reached in the destruction of the environment by man has created a veritable war-situation, and a war-
situation more catastrophic and world-wide than has ever been seen
before. All the nations on earth will have to take the necessary mea-

sures immediately, and, as in wartime, enforce them with all the powers of the state.

Such is the gloomy forecast of reputable scientists. Immediate measures to be taken can, without exception, be summed up in one word: recycling. Our drinking water has become as scarce as that of the pitcher-plant, so we shall have to use it as rationally as does the pitcher-plant. Water must not be treated only when we need it again, but immediately on pouring it away as 'waste'. Plants do not give off dirty water, but only purified water through evaporation. Only through recycling shall we have enough clean water; only thus shall we escape the writing on the wall of the computer prognosis.

Dischidia leaf sacs have an additional, and lesser, function: when they hang perpendicularly from the twigs with their openings upwards, they act as cisterns and food reservoirs, for in this position they can catch rainwater and the products of decomposed insects which have fallen in and drowned.

Many American bromeliads colonize areas similar in climate. They are representatives of the pineapple family (*Bromeliaceae*). They too prefer to grow in the branches of tall trees, where they are exposed equally to burning sun and drying wind, can often expect only very little rainfall, and on the whole must satisfy their need for water from moisture in the air and especially from the frequent night mists. They have acquired a technique different from that of the orchids and milkweeds, since many of them do entirely without roots, while others develop roots only as gripping devices which often have to anchor and hold a considerable weight. Most prefer the direct   90
means of absorbing water, from the air straight into the leaf. For that, of course, some special apparatus is necessary, which the bromeliads possess. It consists of tiny scales that suck moisture directly out of the air.

Aechmea chantinii is a bromeliad familiar to many people   57
because of its distinctively striped leaves. In fact, the stripes are made up of a multitude of round scales in the form of tiny   58
funnels, the points of which are sunk deep into the leaf. Their   59
edges are not attached to the leaf but simply rest on it, and in many cases overlap with one another. The little funnels themselves consist of many single cells, with a diameter of about a hundredth of a millimetre. These microscopic structures are the smallest suction pumps in the world. They are hollow cells

which collapse when dry, but whose walls swell considerably
and become taut as soon as they are moistened. The whole cell
then distends, and a partial vacuum develops in the interior
which operates on the surrounding area as active suction.
Quickly the cell pumps itself full of the available moisture.
From there, the water passes by osmosis into the more highly
concentrated sap in the cells of the leaf's interior. The density
of the suction scales is frequently so great that a considerable
intake of water is possible. Indeed, dry scales will immediately
absorb whole drops of water.

Many of the bromeliads that can be seen growing down from
their host trees like great beards – *Tillandsia usneoides*, for example – are so light in the dry state that they can float. But
when laid on the water, their scales rapidly pump themselves
full so that the plants sink. In their natural environment they
build up such masses of substance from air and moisture alone
that they are used locally in great quantities as a packing
material, under the name of Louisiana moss.

A quite different system of utilizing moisture from the air
has been developed by a number of desert and semi-desert
plants. In order to give an idea of how it works, I will briefly
describe an interesting technique that has been increasingly
applied over the last few years in progressive paint firms:
electrostatic paint application. This makes it possible, using
specially constructed spray guns, to spray finely atomized paint
actually round corners, not along some arbitrarily curved
course, but in such a way that all the particles are attracted
directly to the object to be painted, whether from the front,
the side, or even from behind. The secret is that the paint
particles in the gun have a strong electro-magnetic charge, and
are thus attracted to the object in order to discharge themselves.

60

The manufacturers claim a saving of up to 60 per cent in
paint, which is a considerable technological advance. But for
desert plants there is nothing new here. The situation, slightly
modified in their case, might be described as follows: if it were
feasible to guide the fine water particles contained in the atmosphere towards plants by electrostatic means, as it were
'magnetically', the utilization of the moisture in the air could
be greatly increased. In this instance the gain would be far more
than the industrial saving of 60 per cent, for that estimate is based
on the paint jet directed at the object before the introduction of

25 Plastic resin sheets and panels become much stronger if glass fibre material is incorporated.

26 Strengthening with fibre material also ensures the great stability of many plant cell walls (here *Valonia ventricosa*).

27   Where the stability provided by irregularly arranged glass fibres does not meet the demands of plastic resin construction, so-called twill web or roving web is introduced.

28   Exactly the same kind of reinforcing web is also found in many plant cell walls, as can be seen in this electron micrograph of a cell wall of *Alstonia spathulata*. The enlargement is × 20,000; the mark represents one-thousandth of a millimetre.

30 The areoles, or bunches of thorns, on this pincushion cactus
(*Mammillaria lanata*) are very precisely arranged in spirals.

31a The areole network of the plant in the previous photograph, looked at from the side, resolves into straight lines ...

31b ... which were reproduced in this diagram exactly from the original. This line screen is a construction in accordance with the golden section, or rather with the series of numbers connected with it, and known as the Fibonacci series: 0, 1, 1, 2, 3, 5, 8, 13, 21, 34, 55 ...

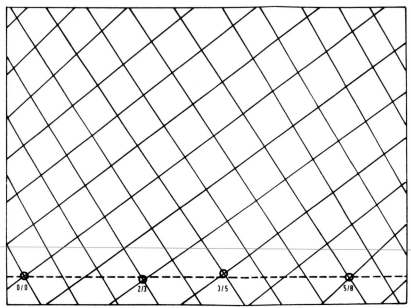

32   Titian's *Bacchus and Ariadne* (in the National Gallery, London) is, like many pictures of both old and modern masters, composed in accordance with the golden section. The grid painted on it is the same as that in the previous illustration.

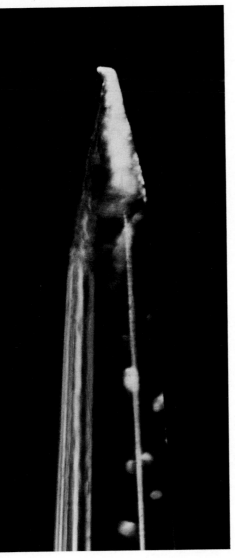

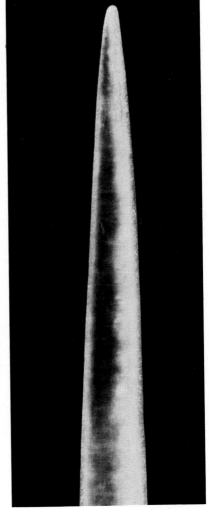

33   The highly magnified point of the precision sewing needle shown here is only 2 mm long. Under the microscope one sees how unevenly the minute tip has been fashioned. Such curved points are very often to be found.

34   Here, at the same magnification as the sewing-needle, we see the extremely precise formation of the point of a cactus spine (*Soehrensia*).

35 Only micro-structures as carefully produced as the point of a record-player sapphire needle can rival the precision of the cactus spine.

36 The floats of the water hyacinth (*Eichhornia crassipes*) from the tropics of South America embody the principle of the pontoon float.

39 The dandelion clock of our childhood is a master of parachute flight.
After ripening, the seeds do not take flight on the first gust that comes
along but, like most airborne plants, wait for a good wind.

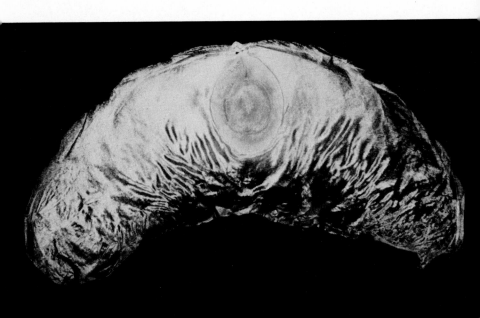

40 Perhaps the most sensational of all plant aircraft, the seed of the
Zanonia macrocarpa has two curved wings 5 cm broad and 7–8 cm long,
giving a total wing span of 14–16 cm. The tissue of the wings is transparent and gleaming, with a texture as elastic as sheet mica.

41 The flying pioneers, Etrich and Wels, made use of the *Zanonia* seed
at the beginning of this century as a design model for their tailless gliders.

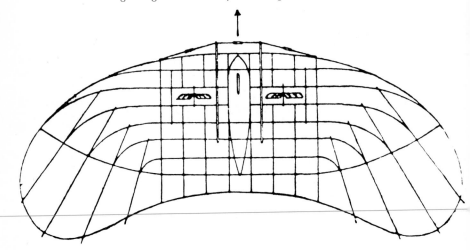

42 Etrich and Wels' first *Zanonia* flying model (1904) had a wing span
of 6 metres and could carry a service load of 25 kg. The second was bigger,
with a wing span of 10 metres, and could carry a disposable load of 70 kg,
but like its predecessor was pilotless. It could glide for about 900 metres.

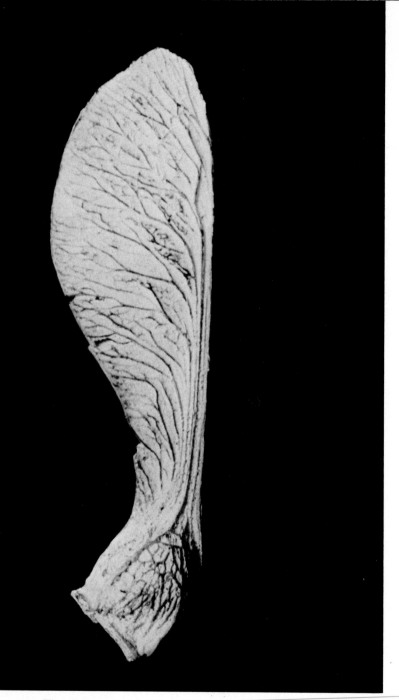

43 The white seed of the Norway maple clearly shows the strutting ribs that make it possible to keep the whole shape very light.

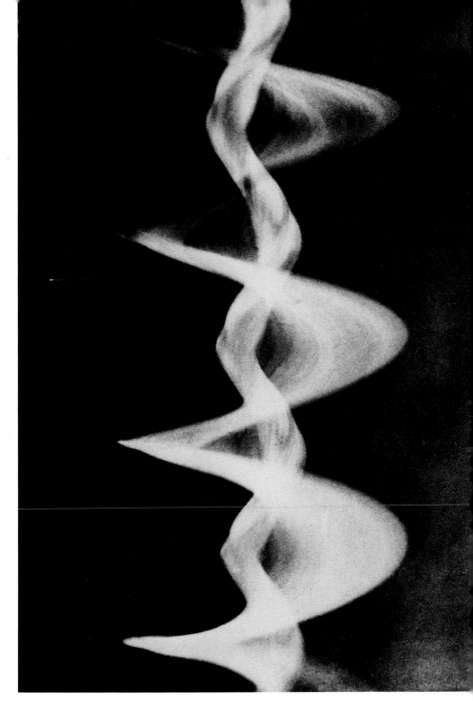

44　In flight the maple seed acts exactly like the propeller blade of a helicopter landing with the engine cut out.

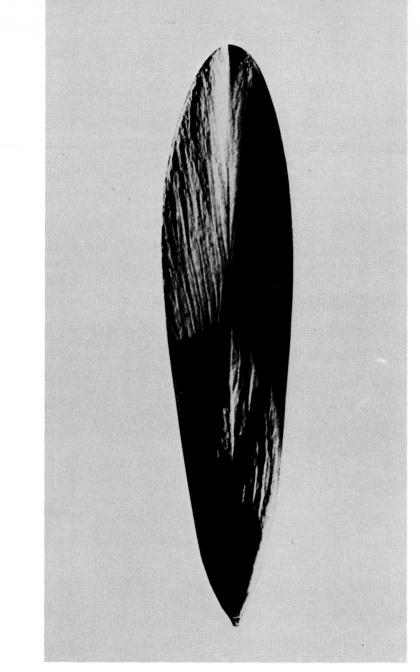

45 Ash trees develop a gyroplane seed that is symmetrical but twisted on its long axis, like an aircraft propeller.

46 The twist of the propeller blade of a sports aircraft can clearly be
seen in this photograph of the end of the blade, which is only apparently
cut on the slant: when the lower end of the blade is square to the
observer, the upper left point is farther forward than the upper right one.

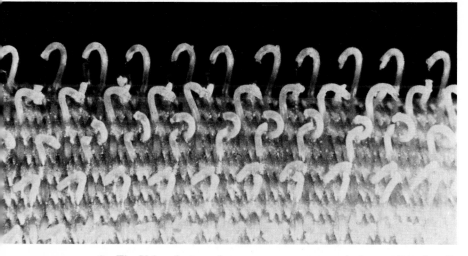

48 The Velcro fastener: into a woven tape are worked a number of small plastic hooks which, on contact with a corresponding velvety tape, catch in the pile and stick firmly; a sharp tug, however, will easily detach them undamaged.

49 The fruit of the burdock, seen at the same magnification as in the previous photograph, shows a very similar structure. The shafts of the hooks are much longer because the animal's coat is stronger and coarser than the material used for the man-made Velcro tape. As for the retaining hooks themselves—are they not a great deal neater and more precise, and thereby better suited to their purpose, than the human product?

the improved system, which is not the case with desert plants. Without the use of an electrostatic system, plants have to depend on the atmospheric moisture that comes into contact with them purely by chance. But they have learned to practise electrostatic 'self-painting' with the mist contents of the air. In contrast to the paint-spraying units, the plants cannot actually give the particles of moisture an electric charge, for these of course are inaccessible. Their solution is to charge themselves. The horny spines and hairs of cacti and other desert-dwellers collect small electric charges from the air, just as an ebonite comb does when drawn through the hair. As the comb then attracts our hair – often we hear a slight crackling noise and even see tiny sparks in the dark – so the charged cactus spines attract droplets of water from the air. In addition to this, they even cause evaporated water to condense. It has not yet been ascertained, as far as I know, how much moisture is gained through this technique, but it must be a very considerable amount. In areas such as the coastal sand deserts of Chile where there are heavy night mists, cacti, which are up to 95 per cent water, can flourish for years without rainfall.

To bring a litre of water at 14°C to the boil, thereby heating it through 86°C, you will need about 0·1 kWh (kilowatt-hour). This is the same amount of energy as a 100-watt electric bulb uses when it burns for one hour. Instead of heating a litre of water by 86°C you could, with the same energy, heat 86 litres of water by 1°C. Conversely, the energy freed when 86 litres of water are cooled by 1°C would provide current for a 100-watt bulb for an hour, provided that the thermal energy could be converted into electricity without loss. But even with some loss, the energy gain would be considerable. If, for instance, it were possible to cool all the water of Lake Constance by only 1°C, about 75 billion kWh of energy would be freed. Even with a poor efficiency rate of 50 per cent in the conversion of thermal into electrical energy, a cooling of the water of Lake Constance by only 4°C spread over a year would still be enough to provide the whole of West Germany with electric power. Nor would the cooling do any harm to the water. On the contrary, Lake Constance and almost all the inland waters of all the world's industrial states urgently need a lowering of temperature. Overheated waste water from industrial plants and power stations has been warming them for many years, causing oxygen deficiency and the resultant death of huge quantities of fish. Rivers no longer purify themselves satisfactorily, and standing waters are becoming covered with algae, and clogged with mud and oil. From this point of view alone the cooling of rivers and lakes is an urgent necessity, if they are not to become irreclaimable cesspools. Why then do we not combine necessity with utility by creating electrical energy from the superfluous heat of our waters? Technically, it is certainly possible.

A successful experiment carried out several years ago by the town of Zurich, which heated public buildings by cooling the water of the river Limmat, proves this, and it has also been done in other countries. Why is this not done on a larger scale? The reasons given are unconvincing: one is that the necessary plant is expensive in relation to the result. But in the very near future every country will have to spend billions on the conservation of its water. Another reason given is that the conventional methods of producing electricity (including nuclear power stations) are technically better developed and therefore more economical, but this only means that we have got used to them and are too lazy to think out other systems seriously and in time.

For the same reason, the entire population of the developed world suffers today from the poisonous exhaust of cars although, had we started developing them early enough, we could now be driving cars with chemical motors whose combustion residues would be liquid and absolutely innocuous. What is worse, leading car manufacturers already have a blueprint for a chemically propelled car, but a tacit agreement forbids the building of these vehicles because the rapid conversion of the production lines would cost too much. Is the price of this economy to be mass death by smog? Should we, from similar dangerously superficial profit motives, prefer the erection of further power stations and nuclear energy installations to the rational exploitation of our waters and the cut-back on the injurious overheating of these waters, simply because coal, oil and uranium, unlike water-heat, can be sold for profit, without considering how long the supplies of coal, oil and natural uranium will last? And because power plants built according to a well-tried formula are cheaper to install and therefore more profitable to sell than thermal pump installations, which would at first entail greater development costs?

Certainly, industrial competition and the laws of the market economy would not allow of any capital investment which, though it might be valuable in the long term, in the short term could lead to the financial ruin of great industries. But the real roots of the failure in long-term planning lie much deeper. Man is not capable of real long-term thinking and systematic action, and never works out fully the possible consequences of his acts. Once a solution has been found to a technical problem, he clings to it without considering alternatives until it has led him to a dead end, by which time it is rather late to look for the way out. Such are the typical consequences of thinking in constructs, unrelated to the laws of development and of feedback to the environment.

We have seen a few modest attempts to make practical use of superfluous water-heat. But how many cheap and well-nigh inexhaustible sources of energy remain quite untapped simply because we have failed to develop technical means of exploiting them methodically and systematically? What quantities of energy could be obtained from the daily variation of temperature between day and night? How much hydraulic power could be freed for conversion to energy by exploiting the relative variations in atmospheric humidity? What unimaginable quantities of energy are available in the form of solar radiation? Even the pull of the moon could be exploited through tidal power stations. However startling it appears, this idea is now decades old, yet so far there is no economically viable station of this kind. Meagre beginnings have been made in the utilization of solar energy, for example in the form of solar batteries feeding to the earth the wireless communications of space-ships and moon stations. Variations in atmospheric humidity and temperature have been exploited to a very modest extent for the automatic operation of greenhouse ventilation windows (a development for which, significantly, we have to thank the botanists among the technologists). But the overall picture is one of complete inactivity.

It is otherwise with the plant: its use of solar energy for the production of organic substances from air and water would be impossible to excel. All the wood and peat, all the coal and tar, oil, natural gas and petrol that man has ever used is nothing but a great reservoir in which plants have trapped solar energy. The significance of sunlight for plants, and their use of ocean currents and wind for propagation purposes, has been discussed. Now we may consider how botanic organisms convert variations in atmospheric humidity directly into useful energy.

In the southern regions of North America – New Mexico, Texas, and Mexico – there grows a small plant called bird's-

61   nest moss (*Selaginella lepidophylla*), which because of its interesting reaction to atmospheric moisture is frequently sold at fairs as a curiosity. In its native desert regions this plant often has to endure droughts that last four years, and it learned long ago to distinguish between damp periods favourable to growth and dry ones hostile to it. The difference between the two is perceived by the plant in the varying degree of humidity. In low atmospheric humidity the bird's-nest moss shrinks in on itself. Vital functions are almost totally suspended. The chlorophyll moves rapidly into the protective interior of the cells, the dry ball takes on a yellowish-brown colour, and the plant looks dead. The spherical form ensures the smallest possible surface area, thus reducing further water loss to a minimum.

When it rains again, after weeks, months or even years, the plant begins to unroll within a few minutes, and in little more than a quarter of an hour is green once again, and spread open

62   on the ground. How does the bird's-nest moss achieve this life-preserving movement? The dryness against which the plant needs to protect itself is used by it in rolling up; the moisture that makes normal growth and life possible again also provides the energy for unfolding. The principle is simplicity itself: on

63   the upper surface of the little branches are tiny scales which absorb water very quickly and easily, swelling considerably in the process. The size of the underside, on the other hand, remains unaffected by the presence or absence of moisture. Thus the rolling and unrolling process is directly dependent on the degree of humidity prevailing at any time. The principle is so reliable that it still functions when the plant has long been dead. The specimen in the illustration has certainly been dead for over fifteen years.

It is not only among mosses that we find such 'resurrection plants', as they are popularly called. The 'rose of Jericho' (*Anastatica hierochuntica*), which grows in many areas from Iran to Morocco, behaves very similarly to the bird's-nest moss. As a mustard plant it belongs biologically to an entirely different family – even to another order – and its home is in another part of the earth, and yet the same environmental problems have caused it to discover the same solution.

This principle of hydraulically induced movement has proved to be the solution to a whole series of other problems, and many seeds and fruits are equipped with a 'hydraulic motor'. The fruits of the cornflower, for instance, can escape from the parent plant by 'crawling' movements. They have a coronet of bristles at one end, with short, stiff thorns pointing forwards. In dry weather the bristles spread out; in damp weather they close up. In this way the fruits are pulled along the ground as the pointed hairs dig into the earth. The fruits of the hare's-foot trefoil (*Trifolium arvense*) creep along in exactly the same way. The fruits (in strict botanical terms 'sham fruits') of the wild oat (*Avena fatua*) have applied the same principle to develop nothing short of a hopping movement.

It is not only for travelling that seeds and fruits employ movement through variations in humidity. Both movement and anchoring function are interestingly combined by the fruits of the old man's beard (*Clematis vitalba*). As old man's beard is a climbing plant, the ripe fruits are often set on the runners high above the ground in the branches of trees. A single curved, feathery plume about an inch long enables them, using just one technical principle, to perform four operations simul- 64
taneously by making use of the moisture in the atmosphere. This hydraulic mechanism, which moves the single hairs and the whole plume in the same process, operates as follows:

1 In rainy weather dissemination of the fruit by wind is highly improbable, for as we know it grows high above the ground and the rain would soon wash it down to the earth. But as it flattens its hairs in the rain, there is scarcely any danger that the fruit will be caught and blown away by the wind.

2 In dry sunny weather, which mostly brings favourable up-currents, the light seed with its hairs fully open becomes a practical flying apparatus. It can now be caught by the wind and carried off.

3 Having landed somewhere on the ground, the fruit begins to move, bending and stretching with the variations in atmos-pheric humidity. As its plumed end easily catches in the grass or in bumpy ground, the seed end then sweeps to and fro, 65
combing the ground for a crack in which it can settle, and which will later afford it enough stability and moisture for growth.

4 If the seed end has found a foothold and the feathery end is still caught somewhere, the alternating hydraulic movements cause the seed to turn on the spot. Its sharp end with the fine 66

barb-bristles is thereby screwed like a drill into the ground and firmly secured.

Thus the feathery plume of the old man's beard is a unique, multi-purpose hydraulic mechanism.

The key to all hydraulic motion, including that discussed in the previous section, is varying hydraulic pressure. In man-made machines the necessary pressure is almost always produced by compressors or pumps. These can be motorized units, such as large hydraulic presses, rivet hammers and power-shovels, or manual pumps like the simple hydraulic automobile jack. The pressures with which such tools operate vary between 70 and 200 atmospheres, which corresponds to the pressure of water in the ocean at depths of 2,100 to 6,000 feet.

These high pressures are necessary because hydraulic machines, even when relatively small, are required to be extremely powerful. They have to operate 100-ton presses, jack up cars, or even raise whole ships. Reliable control of machines, too, or of movable parts, is often achieved hydraulically. Cylinders, high-capacity motors and the flaps of large aircraft belong in this category. Here again the operating pressures go up to 200 atmospheres.

The problems of movement and steering in the plant organism demand very little power in comparison. But they are likewise solved hydraulically, though mostly at much lower pressures. Yet here too we occasionally come upon coefficients equivalent to those of technology. Cell pressure in sugar-beets rises to more than 50 atmospheres, and pressures of 200 atmospheres and more have been measured in many desert plants.

Autonomous plant movements, then, are hydraulically controlled. But does not the production of the necessary hydrostatic pressure presuppose some autonomous pumping movement? In human technology, engineers produce high pressure through compression. Any movable part, for example a piston, compresses the fluid. Such a procedure, using a great deal of energy and necessitating specialized, complicated equipment, would be too uneconomical for plants, which use instead a method structurally simpler and much more effective in terms of energy consumption. They employ osmosis. This method of generating pressure exploits the natural tendency of salts to attract water in which to dissolve, thereby diluting the solution as much as possible. When a dessert-spoonful of cooking salt, soda, sugar or the like, is put into a pan of water, the dissolving crystals at first form a highly concentrated solution, which is then gradually dissipated in the surrounding water until the same concentration is reached throughout the vessel. If the salt is put, not directly into the water, but into a filter bag hanging in the water, three things can occur according to the type of filter. It may be of such a fine material that the water molecules cannot pass through, and then of course nothing happens. Again, the filter pores may be large enough to allow the water molecules to pass through and dissolve the salt inside. And if the pores are also big enough for the dissolved salt particles to pass through them, then we have the same effect as if the filter were not there. After a while the salt solution will have the same strength throughout the pan. A third possibility is to use a filter bag which allows the water to pass in but not the dissolved salt particles to pass out. Chemists and molecular physicists call such filters 'semi-permeable'. The concentrated salt solution that forms within such a semi-permeable filter bag endeavours to dilute itself and to attract more and more water for that purpose. And so more and more water presses into the bag from without, while none of the solution comes out. In this way the bag fills and swells. Inside it pressure builds up, the intensity of which depends on the amount of salt. Quite considerable pressures can be generated in this fashion.

Plant cells are just such little bags filled with salts in solution and surrounded by a semi-permeable membrane. If they get into water, they suck it through their walls, because the salt solution within them aspires to dilute itself further. If a dried-up, soft sugar-beet is put into water, it will very quickly become extremely hard. But if a plant bursting with sap is put, not into pure water, but into a salt solution more concentrated than the liquid in the cells, exactly the opposite occurs. The sap passes out through the cell walls in order to dilute the surrounding liquid. This process can be observed in the preparation of pickled gherkins. A fresh, peeled gherkin quickly loses its firmness in salt water and becomes soft.

Thus plants produce hydrostatic pressure by means of this osmosis. Cell pressure gives stability to plants that do not form wood. If it relaxes, as for instance in cut flowers in a vase, the flowers wilt. Cell pressure is also the driving force in autonomous plant movements. The appropriate joints are there in plenty. They work much like hydraulic articulations in man's

67   technology. In the case of, say, a power shovel, the joint is push-
      ed up and down like a piston by the changing pressure, and is
      directly connected to the movable arm. The same kind of thing
      happens in a plant's joint. If the stem of a mimosa leaf has to
68   move downwards, there is a sudden decrease in pressure on the
      underside of the joint. On the upper side cell pressure is main-
      tained. The downward movement of the stalk is therefore
      practically instantaneous.

      The problem of pressure reduction is solved by the plant's
ability to change the size of the pores in the semi-permeable
osmosis membrane. If the pores become bigger, the salt solution
under pressure can get out of the cells, and the pressure falls.
Within a few minutes the cell walls recover their usual character
and the movement can be repeated.

      Plant movements may be directly controlled by humidity,
but plants can also execute independent movements by changing the osmotic character of the cell walls. The sensitivity of the
mimosa is well known. And indeed the pinnate leaves of the
sensitive plant, as the mimosa is also called, are averse to any
approach. At the slightest touch the delicate pinnae fold up and
fall into an oblique position, the two halves of the leaf close up,
and finally even the leaf-stem sinks. If the irritation is sufficiently
great, the parts of neighbouring leaves react in inverse order.
Much has been written and many have been the conjectures
on the possible significance for the plant of this strange behaviour. The theories of Professor Haberlandt, who has studied
mimosas in their native habitat, seem to me to come nearest to
the truth. He reports:

Later I repeatedly observed the sensitive plant as being one of the
commonest and humblest weeds in Singapore, Java and Ceylon,
although it is a native of Brazil and was imported from there into the
tropics of the Old World. In greenhouses its growth is as a rule more
upright, but in the open air it creeps along the ground, so that at any
shock the irritated leaves can for the most part take shelter round the
stalks, which are armed with thorns. It seems to me that we see
here the main biological significance of this plant's striking reactions:
it is endeavouring to protect itself against grazing animals. Nevertheless I have repeatedly seen the great humped oxen of Singapore
devour branches of *Mimosa pudica*, undisturbed by movements of
irritation and thorn stabs. One can only conclude that, in its original
home, the plant is generally at risk from smaller, more fastidious
animals. Furthermore, it is not impossible that the sudden move-

ments of irritation in the leaves are a means of frightening off harmful insects, which at once lose their foothold whenever they attempt to settle on the leaves. I cannot remember ever having seen these leaves damaged through being eaten by insects. Lastly, there is no doubt that those leaves that have assumed the position of irritation are thereby protected against damage from extremely heavy tropical downpours.

The sensitivity of the leaves is naturally greater as a rule in mimosas growing in the open air under a tropical sky than in the often sickly specimens of our greenhouses. Many a time, when peacefully sketching a tree or painting a landscape, I have suddenly seen a dark gap appear in the thick growth of the mimosas near me, without any visible cause; suddenly some leaves had sunk, and quickly the quivering movements spread in an ever wider circle, accompanied by the soft, scarcely audible rustle of the leaves brushing against one another. The movement spreads because, in the interlaced tangle of twigs, each leaf that sinks touches and thereby irritates another. Only a very strong shock, such as that caused by a wound, communicates itself through the stem from leaf to leaf. (*Op. cit.*, pp. 36–7)

By contrast, the active movement of the stamens of many flowers is quite clear in its aim. When an insect flies onto a flower to gather nectar, the stamens, lightly touched, close over it to unload their pollen, which the forager must then carry on to the next flower.

The Venus fly-trap likewise reacts to a touch with movements of its leaves. But its action is much less innocuous than the mere unloading of pollen. If an insect settles on one of its leaves, within a fraction of a second this leaf, turning on a central hinge, snaps shut like a book. Its edges have long, upturned teeth which interlock with one another and bar every way of escape. The trapped insect is immediately digested by the flesh-eating plant: precipitated plant juices decompose it, and the leaf surfaces absorb the richly nourishing solution. This plant thus illustrates how quickly hydraulic movements can take place in plants when necessary.

It need not always be a touch that stimulates these self-protective, predatory or other reactions in plants. Electrical tension, light rays, variations of temperature and humidity, can all be the source of an irritation that provokes its own set of movements. And all these movements are controlled by cell pressure. The alternate play of light and shadow leads to 'sleep' movements. Temperature and atmospheric humidity control, among other things, the opening and closing of snowdrops and crocuses. Electrical or thermal stimulation can provoke reactions similar to those of the mimosa. The stimuli can thus be very varied. The principle of movement is one and the same: an extremely rational hydraulic propulsion.

CHAPTER NINE

Applied Thermodynamics

Anyone wishing today to protect a building, a caravan, pipes, sensitive apparatus or indeed himself from cold or from excessive heat, will not find it difficult. Specialist firms offer a wide range of insulating materials for every possible use, from asbestos board and plastic foam to double-glazed windows and quilted anoraks. The range is enormous, and the principles applied apparently limitless. On closer inspection, however, all the various materials turn out to work along the same lines.

All insulating devices are based on the low heat conductivity of gases, especially of air. The purpose of the material itself is to diminish the exchange of heat through flow and radiation by dividing up the volume of gas, without increasing conduction to any great extent by creating 'heat bridges'. Insulating materials can be porous (cork, diatomaceous earth, foams), fibrous (mineral wools), or layers of air enclosed in foil (e.g. 'alfol' insulation). These three categories include all known means of thermal insulation. But there is another method of protection which is based, not on insulation, but on reflection. Anyone who has travelled in hot countries in the height of summer in a black car will know how the coachwork of the car reflects the fierce rays of the sun only very weakly. In a white car it could be as much as 10°C cooler.

These four groups – porous, fibrous, foliate and reflecting – exhaust the variety of thermal protection materials. As we have seen, the plant world, in solving its technical problems, generally discovers all the possible solutions, and develops each to perfection. Let us see whether this is so in the sphere of insulation.

When we consider porous materials, it is immediately obvious that man has access to at least one excellent botanical insulator, namely cork. This need not always be as well developed as in the Mediterranean cork-oak which provided the corks for our bottles. All trees form sheaths of cork with their bark, but they are much thinner than in the cork-oak and cannot be exploited industrially. This most important layer of a tree-trunk lies between the heartwood and the bark, only a few centimetres under the surface – in young trunks only a few millimetres. That in itself explains the protective function, both thermal and mechanical, of the cork. I have observed that, in the frequent forest fires of Mediterranean countries, cork-oaks often survive even this great heat more or less unharmed, and produce new growth the next year. This they owe entirely to their insulating sheath, which is several centimetres thick. For insulating power, cork can be ranked with stone wool and glass fibre: burnt diatomaceous earth, a kind of earth consisting of the shells of diatoms which is often used as an insulating material in the building trade, is only as effective as cork if applied three times as thickly.

There are only two kinds of insulating materials that are actually better than cork, by about 25 per cent, and these are synthetic resin foam, and facing foil, which encases a layer of air. The structure of a modern polystyrol resin foam consists of a multitude of very small air-bubbles, of which the biggest measures only one millimetre. The haphazard arrangement and the extremely thin walls of the bubbles mean that those 'bridges' which might conduct heat directly are not a significant factor.

On the whole, therefore, the foam materials are ideal as heat insulators. As far as our technology is concerned, they are a relatively recent discovery of this century. Nature has known them much longer. A natural foam material protects citrus fruits against frost and excessive heat. The cell tissue of these is so distended that it breaks up into thin walls, and thus solves the problem of heat bridges.

Materials with closed pores would not be advisable in cases where, besides insulation, breathing must be catered for. Examples from human technology are wall insulation and warm clothing. These are the chief fields of application for fibrous materials. Glass fibres, mineral wool, asbestos sheets, and textile fibres fall into this category. Plants, too, are familiar with the whole range of insulating fibres, from the short-haired but thick white felt on the petals of the edelweiss to the wool, four or more inches long, in which many mountain cacti wrap themselves; from the soft silvery coat of the pussy-willow catkins which often appear in the winter, to the crowding together of innumerable thin shoots in many cushion plants. Amost always the felt, the bristles, the hair and other fibres are in addition white or silvery, and thus offer further protection through reflection.

Finally foil, especially aluminium foil, has a definite role to play in the technological fight against heat and cold. Creased up, or laid close together in parallel sheets, foil insulates well. Plants likewise know the principles of creased and parallel foil, as well as a whole series of mixed forms between these two extremes. The covering of several thicknesses of scales in which sensitive buds wrap themselves might best be compared with the parallel-foil insulation, while the dried-up, shrunken epidermis of the 'living stones' (*Conophytum* species) is more representative 71 of the creasing principle. Alpine plants often protect their living shoots with a multi-layered wrapper of dead leaf-parts, which the experts have aptly, if facetiously, dubbed 'straw tunics'.

Anyone who knows the tropics, the desert or the jungle knows also how scorching and inimical to life the equatorial sun can be. Everything seeks coolness and shade, and yet there is no shade. Plants of those regions must create it themselves with a thick covering of silvery-white scales, as do the little desert roses (various species of *Anacampseros*) on the glaring gneiss and quartz fields of South African desert areas; or with a long white coat of hair, like the 'living snow' (*Tephrocactus floccosus*) which forms whole compact fields in South American mountain deserts; or simply by some appropriate shape. Many cacti have developed deeply ribbed forms which ensure that, when 74 the light is not directly overhead, the greater part of the plant is in its own shadow. Vertical rays fall only on the tops of the ridges or the top of the sphere.

The spherical form, as compared with European foliage, represents an excellent adaptation to fierce sunlight, since one half of a globe, like the earth itself, is always in the shade. It is 72 no accident that domed buildings are common in the East. 73

It is very interesting to see that quite different plant families, independently of one another, have solved the same problem in exactly the same way, and therefore look very much alike. A 74, 75 cactus (*Neoraimondia gigantea*) of North Peru, for instance, and a spurge (*Euphorbia canariensis*) from the Canary Islands may appear almost identical as a result of the same climatic conditions and corresponding environmental problems.

The most sensitive zone of the cactus's body, the growing tissue at the top, would certainly be also the area most exposed to the rays of the sun, were it not protected by a whole phalanx of shade-giving thorns. Often these are arranged like the spokes

of a wheel to form dense, effective sunshades, which at the same
time let in plenty of air.

If spherical and column-shaped plants, whose form protects
them from both excessive light and excessive evaporation, are
at home in the deserts of the tropics, in the equatorial rain
forests there are plants which, though they can afford the luxury
of large evaporating surfaces, need just as much protection as
do the desert dwellers against the intense sunlight of their
latitudes. The typical vegetation of these regions consists of
trees. But even externally they differ greatly from the trees of
our woods. The branch and twig growth of tropical trees strikes
a European eye as strange and haphazard, often downright
ugly. The leaves hang down apparently limply, or look as if
they had been distributed at random over the tree: entire
branches are almost bare, while at their extremities the leaves
grow all together in tight bunches. The silvery boughs seen
among dark green foliage are tortuously twisted or make
sudden, right-angled turns. The foliage has none of the transparency that goes with the mild light of our woods. Tropical
leaves are thick, hard, dark, and almost completely opaque, as
though they were made of green-painted tin. Their surface is
often smooth and waxy, and reflects the sun's rays. Innumerable
gleams of light reflecting from the foliage glitter all the more
brightly against the dark background, irritating the eye accustomed to the woods of temperate regions. Anyone who has tried
to photograph a rubber-tree in full sunlight will have some idea
of the contrasts in brightness produced by tropical trees growing wild.

What causes the striking difference between our home
forests and those of the tropics? It is the light. The mild light
of temperate latitudes must be fully exploited. The leaves lie
horizontally in order to catch as much of it as they can. They
distribute themselves over the whole tree so as not to overshadow one another. Hence the even, methodical growth of
branches and twigs. When the direct sunlight becomes too
warm, as occasionally happens, the foliage of our trees protects
itself by increased evaporation. Evaporation has a cooling effect.
Light in the tropics, on the other hand, is dazzling and intense.
The tropical sun does not warm, it heats. Overheating can be
avoided only in the shade, and the plant must create that for
itself. So it is no wonder that the leaves crowd close together at
the ends of many branches, to give shade to one another, nor

that the foliage often hangs perpendicularly from the stems so that whole trees, to our eyes, look as if they were withered. This position means that the rays of the sun touch the leaves only very obliquely. And the smooth, gleaming surface reflects as much light and radiated heat as possible.

An admirable example of a plant creating its own shade is the traveller's tree (*Ravenala madagascariensis*), a native, as the botanical name indicates, of Madagascar. At the top of a slender trunk several yards high, it bears one single enormous fan of leaves, symmetrically spread on both sides. The youngest leaves are in the middle, pointing vertically upwards, so that the sun's rays meet them only at a very acute angle. The older leaves on both sides lean further and further down till at last they are in an almost horizontal position. But they are so tightly packed together that they still give shade to one another. Writers have stated that the giant leaf fan is aligned due east and west, whereby of course mutual shading by the leaves would be most complete. This orientation is said to have earned the tree the name of 'traveller's tree'.

But tropical flora cannot win the fight against light and heat radiation by appropriate leaf positioning alone. A real problem of light and shade arises from the strong shadowing effect of the foliage. In the shadow of the giant jungle trees everything presses towards the light, only to protect itself against the light once it has been reached.

Humans also are plagued by the same problem, especially in the summer months. Light offices with big plate-glass windows are favoured places of work, but anyone who spends eight hours in such a place on a hot summer's day soon appreciates the advantages of Venetian blinds in keeping the glaring light away from the desk and tempering the heat. But blinds need a lot of attention. In the early morning they are not really needed, so they are pulled up. Around 10 o'clock, when the sun first falls on the desk, it is enough to let them down half-way with the slats half open. An hour later they are let down completely with the slats still half open. Between 1 and 2 o'clock the midday sun shines full on the office window, so the slats are closed. About 3 o'clock the sky may become overcast, with heavy clouds heralding an approaching thunderstorm. It is now too dark to work in the office, so the Venetian blinds are drawn up again. An hour later the sun is once more shining, and the Venetian blinds are let down. No wonder the roller-blind industry talks of optimal comfort when it offers, as it has done lately, a shading device for big office windows controlled automatically by sunlight.

But expensive, modern comfort and technological advance for us is a matter of course for many plants. Leguminous plants (peas, beans, peanuts etc.), which are commonly found among the trees and shrubs of the tropics, have adapted themselves with particular skill to alternations of light and darkness, heat and cold. Like the latest in Venetian blinds they automatically regulate the intensity of the heat and light by infinite variations in leaf position, from the horizontal to the vertical. When the degree of light suits them, they spread their pinnate leaves flat out; in fierce sunlight they hold them straight up or lay them together in pairs. In the dark they behave in the same way, thereby avoiding an excessive loss of heat, just as they avoided overheating in bright sunlight. No such protection against the sun is required by plants in our latitudes. But the French bean or the wood-sorrel, which is common in damp woods, may be observed to change the position of its leaves during the night in order to prevent heat loss.

A few years ago Zeiss, the world's leading optical glass firm, developed, under the trade name 'Umbramatic', sunglasses that adapt themselves to any intensity of light. In the shade and indoors the glass is almost as clear as water. With increasing brightness it becomes darker and darker, until in brilliant light it reaches its deepest brown. These glasses are a supreme example of industrial development. Their manufacture is costly and complicated, and their price is correspondingly high. When the 'umbramatic' technique came on the market, an oculist assured me that nothing so ingenious had ever been seen before. He was mistaken. Any green plant could have been the model for those glasses, for green plants have long practised the same technique – automatic brown colouring regulated by sunlight – as a matter of course. You can observe it everywhere in our flora. Leaves and shoots that grow in the shade are yellowish-green or a delicate light green. In places with more light they are brownish or tinged with red, and in very sunny places they often become almost dark brown. Shading pigments appear on the surface of the leaf which do the work of sunglasses for the sensitive chlorophyll – and sunglasses of a very expensive kind too, for the shading is automatic and infinitely variable.

50   The complete seed, 8 mm long, of the bur marigold (*Bidens ceruleus*). The four teeth, equipped with barbed hooks, measure only 3 mm.

51 Under the microscope we can see how precisely the 'harpoons' of the bur marigold are made. Each harpoon is about 3 mm long; the section shown here is 2 mm.

52a and b These bone harpoon heads of the palaeolithic period remind one vividly of the plant micro-structure in the previous photograph. The aim is the same: immediate anchoring with the briefest of impacts.

53 The scene in many places six months after the catastrophe of November 1972 was still much as it appears in this photograph of the Upper Spessart. The picture incidentally illustrates another aspect of storm damage: the fallen trunks are for the most part so shattered and splintered that their timber is worthless.

54 A notice on a nature trail pointing out that storm damage is to be avoided by mixed cultures, whereas planting a single variety—in this case the fir—is risky. But however praiseworthy its intention may have been, the notice is today a macabre reminder of man's unwillingness to learn. Newly planted firs may be observed growing all around it.

55 The remarkable leaves of the pitcher-plant *Dischidia rafflesiana*. From
the outside flat and yellow-green, they are hollow inside and have a
small opening (with the edges well turned in) at the point where they
are attached to the stem of the liana.

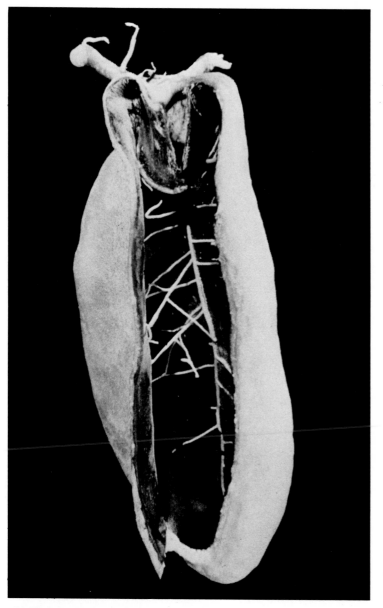

56 The slit pitcher-leaf shows the roots that grow into the dark interior
of the leaf sacs, there to ramify abundantly. Water evaporating through
the leaves condenses and is absorbed again by the roots. A more
economical form of water conservation it would be impossible to
imagine.

57   A favourite indoor plant, *Aechmea chantinii*, absorbs water from the air directly on to its white-striped leaves.

58   Under the magnifying glass the stripes resolve themselves into a multitude of small white circles (the average width of the stripe is 11 mm).

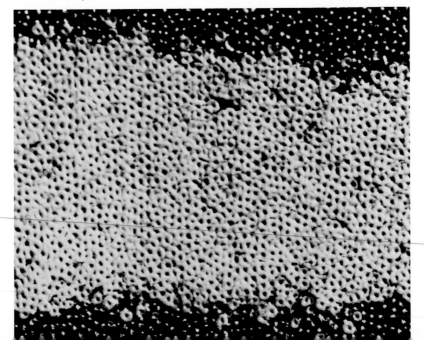

59 Only the microscope shows the true nature of the white stripes: they are composed of many hundreds of tiny, funnel-shaped suction scales, not more than 0.25 mm in size, and consisting in their turn of agglomerations of microscopic spongy cells. These can clearly be seen round the central depressions of the scales, where they are particularly large.

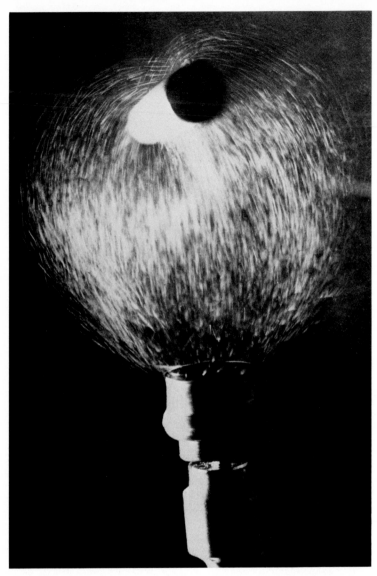

60   Many desert plants attract moisture from the air by electrostatic
'self-charging'. The method is similar to that of the electrostatic spray
gun shown here, where the paint mist does not fly in a straight line but
travels on magnetic paths.

61   The bird's-nest moss of Mexico contracts to a ball of straw in time of drought as a means of protection against drying up.

62   In damp air, on the other hand, the moss expands within a few minutes into a flat rosette. The movement is hydraulically induced. Scales on the topside of the branches expand as they absorb water, while the bare underside remains the same, thus causing the plant to open.

63   Seen under a microscope, the hygroscopic scales form a dense covering over the fine branches of bird's-nest moss.

64 The seed of old man's beard reacts markedly to varying atmospheric humidity. Here, the same fruit has been photographed four times on the same piece of film at intervals of a few minutes. In the position furthest to the right, it is still dripping wet, and the hairs are plastered against the stem of the plume. As it starts drying, the hairs soon begin to spread out and the curve of the plume contracts. The position furthest to the left corresponds to the curve of the plume in very dry weather.

65 Through its hydraulic propulsion the seed can execute clearly defined searching movements to find a suitable crevice in the ground where it can germinate.

66 The head of the seed is pointed like a gimlet, and barbed: the movements of the plume enable it to anchor in any crevice.

67 The joint of this power shovel works hydraulically. The movement is produced through variations of hydrostatic pressure in the cylinder.

68   This double-exposure photograph shows how the hydraulic joints of the mimosa work. The white arrow indicates the movement which the leaf-stalk executed a few seconds after the leaf was touched. Immediately the pinnae folded up. Where the leaf-stalks grow out of the plant stem a thickening can clearly be seen; this is the 'hydraulic joint'.

Cacti are often mentioned in this book. One reason may be that this remarkable plant family has fascinated me for almost twenty years, so that I have studied it with great interest, but the main reason surely is that the native environment of cacti often demands from these bizarre columns, cylinders, spheres and coils, a tenacious will to live that results in the utmost extremes of plant adaptation.

A great number of the most comical cacti are native to the dry plateaux of Mexico. Some of them have developed a style of wintering most unusual in plants. When the Mexican ground-squirrels, relations of the Alpine marmots, crawl into the earth at the first signs of the dry season, and sleep through six or seven dry, cool winter months under its protection, the remarkable *Lophophora*, or mescal cactus, from which the drug mescalin is extracted, likewise sinks into the earth to hibernate. There it 78 keeps warm, sheltered from the cold drying winds that sweep over the uplands. With the first onset of the rains in February or March, the grey-green leathery ball crawls out of its earth nest into the light of day.

The principle of burying oneself in winter is surprisingly simple. The mescal cactus, like a whole series of other species, is anchored in the ground by a long thick tuber. With the onset of the dry season in autumn the plant shrinks from the top down, often losing as much as half its volume. In the spring the shrivelled body quickly sucks up all available moisture, expands within a few days to its former size, continues its growth in the heat and light, and soon comes into bloom.

CHAPTER TEN

Communications:
from Semaphore to Computer

Plants are masters of constructional statics; land, water and air transport; hydraulics and thermodynamics; the use of light energy; the measurement of time; and the principles of chemistry. They have outstanding achievements to their credit in every one of these domains. Anyone who looks down the index of an engineers' handbook, or scans the list of departments in a technical college, will perceive that the plant capabilities already described in this book cover a large part of human technology. The principal subjects still remaining are the technologies of measurement and of communications. Both of these are supposed to help man to come to terms with his environment, to learn facts, and to express himself intelligibly. These are also important problems for plants. What solutions have they found for them?

By communications we understand in the narrow sense the relaying, in the wider sense the acquisition, interpretation and storage, of information. Plants are confronted by such tasks when they are working together with other living creatures. This is especially the case with pollination by insects and other animals. In order that the insects may move from flower to flower with the greatest possible gain to both sides (a rich supply of nectar for the insects, intensive pollination for the plants), rationalizing measures must be undertaken which can only be effected through 'communications engineering'. To avoid vain searches, the plants must set up signals which give information on their flowers. In order that insects shall not fly to flowers that are not as yet sexually mature, and therefore can neither be pollinated nor offer anything to the insect, such flowers must be marked. And in order that flowers already pollinated and robbed of their nectar should likewise not receive unnecessary visits, another suitable signal is called for. The signals should be varied in accordance with the nature of the pollinating animals. Bees or beetles need to be solicited in a different style from flies, and flies differently from humming-birds or bats. A flower constructed exclusively for bird pollination should, if possible, attract no insects. Finally, optical signals visible at a distance commend themselves in territory easily surveyed, but in country not thus open to view other methods of attraction and communication must be found.

The catalogue of measures taken reads like the report of an industrial consultant. If the list of problems is long, the number of solutions is so considerable that until now it has been possible to investigate only a fraction of what plants have developed in this sphere. Yet already whole books have been written on the subject. Here as always we must content ourselves with a few examples.

Visual signals are the most frequent and the most important lure for pollinators. 'If the *Musaenda* flower's means of attraction had not long ago been compared to the putting out of flags, the comparison would have occurred to the most prosaic observer,' wrote the botanist, Professor Haberlandt, at the end of the last century, in his description of these Malay climbing plants.

The little five-piece corolla is obviously not sufficient to attract the visiting butterfly, so one of the normally unobtrusive petal-lobes develops into a great, upright, milk-white petal with a gleaming surface, 2 by 4 inches, clearly visible from a distance. Associated with the flag-like form is the 'traffic-light' effect of colour. The white or pale yellow colour is no accident, for butterflies are particularly susceptible to light, clear colours. But if flies are to be attracted, grimy flesh colours will predominate, dull dark-brown or reddish-black shades unpleasing to the human eye. Flowers pollinated by flies not infrequently exude a strong odour of carrion, or smell abominably of rotten flesh, decayed albumen or stale cheese. The colour, the flecked markings, the carrion smell and the often considerable size of blooms visited by flies (the parasitic plant *Rafflesia* develops probably the largest blooms in the world, over a yard in diameter) simulate very convincingly blood-encrusted, decaying carrion, and therefore promise a luxurious meal.

In addition to these plants which use smell signals, there are others which display the types of colour and form most likely to attract the visitor. Let us consider colour for a moment. Bright red flowers are particularly interesting. Only a few insect-attracting blooms are entirely red, for insects cannot recognize a primary red. In fact their capacity for seeing colours has been compared to red-green colour-blindness. Such blooms only attract insects if the red is mixed with other colours, and indeed until now no flower of a pure red colour has been found to be visited by insects. This colour *is* found in flowers, but then the visitors will not be bees, flies, beetles or butterflies, but birds, for example the humming-birds of America or the equally tiny honey-birds of the Eastern tropics. The eyes of these creatures are particularly sensitive to it, and a bright primary red is therefore the colour best suited to attract them at a distance.

A bright colour is also appropriate for attracting the herbivorous bats, which fly about in the evening when the light is failing. Thus *Freycinetia*, a tropical liana adapted to pollination by flying foxes, has bright red petals which indicate to the bats the presence of succulent food.

Flowers which, by specific signals, attract birds for pollination are of course structurally prepared for these visits. The sweet nectar at the base of their trumpets is accessible only to a bird's beak and tongue. But humming-birds often try to reach the delicacy by pecking off the blooms at this point. Naturally this does not suit the plant, which in return for its nectar expects pollination. A South American shrub, *Jochroma macrocalyx*, outwits the humming-bird with complete success. Should it start pecking at the calyxes from underneath, a jet of water greets it. A sadder and wiser bird will in future refrain from honey-stealing, and will use the conventional method in accordance with the plant's wishes.

Insects and birds naturally do not land on blooms that are still closed, and thus avoid a wasted visit. But many plants have open flowers that have not developed any nectar, because they are not yet sexually mature. Any attempt at pollination would be wasted on them. Although we have yet to discover how they do it, such flowers signal clearly that they have nothing to offer, so that visitors are spared the trouble of looking there for nectar, and instead fly straight on to the next flower. The sign is not visible to the human eye, but clear enough to convey exact information on the possible yield of nectar to the insect or the bird, which thus avoids superfluous visits and has time to enter more mature flowers – a gain for pollination. In the same way, plants also indicate pollinated blooms already robbed of their nectar. Mostly this is done by a rapid, striking change of colour, or by immediate withering. Orchid growers know what they are doing when they carefully protect the splendour of their greenhouse blooms from insect visits. Swift withering would inevitably follow pollination.

Besides visual signs, plants emit odours which are effective even when thick bushes prevent a flower from being seen at a distance. In general, adaptation to the type of visitor is as flexible here as in the case of visual signs. Flies have an excellent sense of smell for a few quite definite odours, closely connected with their way of life, but are completely unaware of all other smells. The flowers that wish to attract flies therefore, as we have seen, stink of carrion: no inconsiderable chemical achievement, since this involves imitating odours ordinarily produced by quite different substances. On the other hand, bees apparently have exactly the same reaction to smells as do human beings. No wonder, then, that the flowers they visit also smell agreeable to us. Many butterflies and beetles are masters in the art of detecting smells. Substances which the nose of man cannot detect, even in concentrated form, they perceive when diluted to less than half a millionth of a milligram in a cubic yard of air. Thus they can often smell a single bloom at a distance of several kilometres, and fly directly to it. No walkie-talkie could convey information over such distances with so little expenditure of material and energy. From what distances then must the aroma of whole hedges or trees in bloom attract insects?

The arum, which flowers in spring in low-lying, damp deciduous woods, is trebly equipped with effective signals. Its large petal, rolled into a tube, is the familiar flag, visible from a distance. Anyone who bends over it will be struck by the fairly penetrating, disagreeable carrion smell which informs flies of the nourishment in store. But in addition to these signals the arum flower, like many tropical water-lilies, attracts through heat rays, to which the tiny midges and flies that visit it are particularly receptive, above all in April when the days are still often rather cool. Through intensive exhalation of stored energy the flower-spike acts like a radiator, warming the lower funnel up to 40°C above the surrounding temperature. Thus one of the attractions of this plant is a warm room in which the insects gladly stay, particularly as it offers nectar too. Once in the funnel, the flies and midges are systematically put to use by the plant. Having crawled in through a ring of fine hairs which only bend one way, they are prevented from getting out again until they have pollinated the female flowers at the bottom of the funnel with the pollen they have brought with them. The male flowers then open above them and sprinkle them with fresh pollen, and only then do the hairs at the entrance shrink back and leave the exit open. The arum therefore practises, besides threefold signalling, an entirely different type of communications technology: data processing. Its pollination mechanism functions like a computer-regulated industrial production line.

One of the newest and proudest products of the communications industry is the computer. Calculations which would take a mathematician six months can be done in a matter of seconds with its aid. It can store millions of data units in the smallest of spaces and recall them at any time. Scientific calculations, the keeping of bank accounts, high-speed airline bookings, interpretation of statistics, the processing of opinion polls and the counting of the vote, these are just a few of the almost infinite functions of electronic calculators.

Their fundamental principle is as simple as you could imagine. They can distinguish between nothing but yes and no, larger and smaller, like and unlike, so they can count only from 0 to 1. They must resolve all larger numbers into a sequence of zeros and ones. The number 3, for example, in computer language is 00011, the number 13 is 01101, and the number 22 is 10110. In order to note any one of these three numbers, therefore, the computer needs in each case five so-called storage places. For the number 22, it would go like this: the first storage place notes a 1, the second a 0, the third and fourth each a 1, and the fifth again a 0. Because a computer must often store up a great many numbers, the single storage places naturally cannot be big. Until recently these were tiny magnetic ceramic rings, threaded at the intersection points of a net of wires. Each of these storage rings was capable of storing the distinction 0/1, according to whether it was magnetized or not.

79

A core-store, that is a constructional element comprising many such wire meshes, could contain over 50,000 or even 100,000 of these tiny magnetic rings, and accordingly as many 0/1 distinctions.

Computer technology has made rapid progress during the last few years. Electronic computers have become steadily smaller and faster. Today they work with extremely small magnetic storage rings, so tiny that a human hair could barely be forced through the centre. Their external diameter is one-fifth of a millimetre. Carefully threaded on gossamer-like electric wire, each one effects the electronic notation of the 0/1

80

distinction. But the industry, still not satisfied with this advance, has now developed even smaller 'memory cells'. The finest electrical wires and silicon elements can be fitted onto the plate with minute precision, forming in all 664 switching elements which so combine that the entire structure can note 64 0/1 distinctions. The 24 rings shown in the picture by way of comparison, 1·2 millimetres in external diameter, as used in earlier computers, could each store one such differentiation.

Computer construction therefore owes a great deal to the development of micro-electronics, which was speeded up by the demands of space travel. The great unwieldy installations have become considerably smaller in the last few years. Nevertheless, a complete electronic computer installation, consisting not only of the central storage but of input and output units, additional stores and other apparatus, takes up a fair amount of space even today.   81

You might imagine that man is clearly the superior of nature in the construction of data storage units, and that to get any more information into a tiny space would be inconceivable. But to test this supposition, let us observe the data storage units of plants. We come across them every day, if not, as town-dwellers, in nature, then at least in the grains of our wholemeal bread or the poppy-seeds on our rolls: in short, in the seeds of plants. All the specifications as to the appearance of the plant which will grow from it are stored within the seed. Size, colour and shape of leaves and flowers, behaviour in heat and cold, light and shade, in acid or calcareous soil, in drought or downpour, are clearly determined in it.

Let us try to discover how many 0/1 differentiations would be necessary in order to programme in a computer only the colour of the flower of a plant. Naturally this is not immediately possible, but a rough estimate will suffice to show the size of the problem. We know from colour television that all possible shades can be mixed from the three base colours, blue, green and red, through varying the amounts of each in the mixture. For the sake of simplicity I will not assume an infinitely variable scale in the base colours, but gradations of 1 per cent. In a mixed colour, therefore, the blue portion will be 1 per cent, 2 per cent, 3 per cent, and so on up to 100 per cent. The same will be true of green and red. Thus, crudely simplifying, we still have 5,151 different shades. 13 storage elements are needed in the computer to store any one of these colour combinations.

10 more elements will certainly have to be added to contain the information that this is a colour that is being registered. The specification could just as well be one of size, form, hardness, or any of a thousand other features.

Thus for the colour of the flower 23 elements in all are necessary. Then we have to determine where the colour is to appear: in the root, stem, leaves, fruit or flower. At this point it becomes very difficult to say how many 0/1 differentiations are necessary for the information. Let us assume, however, that only about two dozen kinds of tissue are in question with a plant, which makes 5 further 0/1 storage elements. Thus we have a total of 28 elements so far for the colour of the flower. Let us take a flower of three colours, as typified for example by the common daisy with its yellow centre and white, pink-tipped petals: the number of storage elements rises to $3 \times 28 = 84$. To this must be added data on changes of colour, whether caused by the passage of time between the opening of the bud and the withering of the flower, or by variations in light intensity or temperatures, different minerals in the ground, and so on. Let us restrict ourselves to these four possible influences and assume that, according to its extent, each might extend our 5,151-phased colour scale by a maximum of 500 phases. With four influencing factors and three different flower colours we shall have to programme a further $3 \times 4 \times 500 = 6,000$ variants, which makes another 13 storage elements. But for every influencing factor we have also to determine which of perhaps a thousand possible physical or chemical factors is involved. This gives us a further 40 storage elements per colour, 120 in all. Thus the sum of storage elements required for flower colour is 217. These would establish one of the simplest of the data for the future plant in the computer.

If you have some knoweldge of data processing, you might like to try to programme in the same way the other plant qualities. The task will drive you to despair. First there would be the other colours of the plant, those of the roots, stem, branches, twigs and leaves (including thorns and hairs), fruits, seeds and so on. Then we have quantity specifications for all these organs. Next come details of surface structure, hardness, flexibility, heat conductivity, translucency, and so on, of all the parts of the plant. The mathematical encoding of the outward form alone of a tree – the way in which its roots, branches and twigs divide, the relation of the diameter of the trunk to that of the branches, data on the exact geometric shape of the leaves, buds, blossoms and fruits, particulars of the bark – would entail several million o/1 notations. After the outer structure there is the inner – the disposition of the different kinds of tissue in the plant, the form, size and arrangement of the cells in the duct system, in wood, foliage, and so on. The chemical qualities of the cell sap follow. Finally, exact data are to be stored for the growth processes through the life of the plant, for its behaviour in every conceivable environmental situation, for the manner and time of its propagation.

The storage capacity of a large modern computer would scarcely suffice for all the plant data together. A plant keeps the whole of this information in one tiny seed; in the case of many tropical pineapple plants (for example *Pitcairnia maidifolia*), no fewer than 25,000 such seed grains are needed to make up one gram.

Still smaller are the spores of fungi, yet they too contain complete information on the structure and behaviour of the fungus from which they come. If such a spore had been photographed with the magnetic ring in the illustration, and the photograph then enlarged until the spore appeared the same size as the ring is shown in the book, the resulting enlargement would measure about 11′ × 15′9″. Compare that with the floor area of your living room! The spore measures only five thousandths of a millimetre. In comparison with the measurements of this high-capacity data-store, the elements of the most modern micro-electronics are like a volcano compared to a match. And this is by no means an exaggeration: a volcano of 9,000 feet is 70,000 times as high as a match is long. A very small computer room only 15 feet in length is about a million times the size of a fungus spore. In addition to which a fungus spore is far more than a mere data store: it is at the same time the reservoir of material from which the new fungus plant will develop.

But anyone who imagines that the huge computer must be considerably less vulnerable than the tiny data store of the plant, is mistaken. The computer, to work properly, needs dust-free air and constant temperature and humidity levels. Many plant seeds and most fungus spores tolerate anything from extreme degrees of frost to the boiling-point of water, from absolute drought to constant damp, from dust-free air to desert sand: anything, in fact, that the environment may choose to inflict on them.

CHAPTER ELEVEN

Achievements in Measurement Technique

Adjustment to an environment presupposes above all a knowledge of that environment. For this purpose human beings employ techniques of measurement in addition to their own senses. With the aid of these techniques we also get a clear picture of things that we would apprehend only imperfectly or not at all without instruments. The detection of poisonous substances in the air we breathe or the water we drink; decisions as to the right degree of lighting for places of work or the exact exposure for sensitive films; the precise determination of tiny quantities of a substance, or the exact measurement of the moisture content of well-seasoned rare woods in the construction of musical instruments: these are only a few of the infinite number of tasks that would be beyond our capacities if we did not possess highly developed, sensitive measuring equipment.

Living in a manner suited to one's environment means, first, knowing that environment and then acting on the knowledge. To put it another way, an organism that lives in harmony with its environment is certain to have a good knowledge of it. The behaviour of plants is in harmony with their environment. They are in no way inferior to humans in adaptability to their habitat: on the contrary, in many respects they are superior. Does this mean that they know their environment better than we can, with all our technical aids? Generally speaking, the answer is no. Man with his radio-telescopes has succeeded in picking up electro-magnetic waves from extremely distant stars, and even from some which exploded and vanished long ago. He can register the seismographic waves of an explosion taking place many thousands of miles away. He can determine the precise insolation of Mars or any other planet. But what does he gain from these extraordinary achievements? They do not improve the quality of his life. The plant has learned to do none of these things. Why should it? From the point of view of a purely utilitarian system, the acquisition of such skills must appear an uneconomic, and therefore not very intelligent, digression from real living. Comprehension of the environment does not, for the plant, include knowledge of any distant stars; but it can include, for example, the exact registration of the course of the moon, from which to deduce a suitable pattern of behaviour when living in tidal zones. Functional measurements of this kind include the exact evaluation of light stimuli, gravitational impulses, humidity levels, chemical substances or contact stimuli. Such measurements are directly connected, for plants, with the business of living, and the plant is as well versed in them as man, perhaps even better than man with all his expensive methods of precision measurement.

I have already mentioned the inconceivable sensitivity of growing root-tips or unicellular plants to chemical substances: a fern's sex cell can detect and recognize 0·000 000 028 milligram of malic acid. I have also pointed out that bacteria are better than any chemo-technical analysis apparatus in the detection of minute quantities of oxygen. The transmission of an irritant stimulus in a plant is probably also chemically activated. For this, too, extremely small quantities of active substances suffice. An oxyacid from the compressed sap of the mimosa diluted in the proportion of 1 : 100 million still produces a definite reaction in the plant. The dilution corresponds to 25 drops in a swimming pool 15′ by 60′ and 4 ′6″ deep. This too would be beyond the capabilities of chemical analysis equipment.

The precision achievements of plants in the field of time measurement (which will be discussed again later) would be enough to merit the designation 'chronometer', which is bestowed as a distinction on particularly efficient timepieces by the governing body of the Swiss watch-making industry.

For climbing plants, the evaluation of touch stimuli is of vital importance. When their tendrils touch something in searching for a hold, they must recognize that they have done so and respond at once with a retaining movement. The tactile sensitivity of such special climbing organs far surpasses the human sense of touch; even the pharmacist's scales and the highly sensitive weighing equipment of the laboratory are outclassed. You can measure quantities of 0·01 milligram with a good chemical balance. But a tendril reacts within a few seconds to the touch of a woollen thread weighing only 0·000 25 milligram bending so quickly that the movement can be detected by the naked eye. Moreover, even in the case of such a light touch, the plant, unlike any technological apparatus for tactile measurement, can actually distinguish between different materials. Falling drops of water, for instance, which should not produce any movement of the tendril in their direction, or a glass rod, which is too smooth to be gripped, convey no stimulus.

Equally amazing is the plant's ability to perceive the smallest of luminous intensities. A 25-watt lamp in absolutely clear air and in complete darkness can be detected and located by the tips of a sweet-pea seedling (*Vicia villosa*), which are extremely sensitive to light, at a distance of 19 miles; a 100-watt lamp at 44 miles. There is certainly no very great difficulty in detecting a 100-watt lamp at a distance of 44 miles. Even a candle can be perceived at a distance of 17,500 miles with astronomical telescopes and appropriate luxmeters. (It would correspond to a star of the 23rd magnitude.) But such measuring methods are based on the magnification by a strong telescope of the source of light, and hence of the observed luminous intensity, to a degree perceptible to good luxmeters. If the light is not coming from a point source, the measurement technique breaks down. In such circumstances it could detect only with the greatest difficulty luminous values which, over a longish period, are still registered by the sweet-pea seedling. An extremely sensitive measuring system equal to this task would unquestionably be destroyed on the spot if direct sunlight fell on it, for direct sunlight is four million million times as bright. And yet that efficient optical measuring system, the plant, survives unscathed the tremendous variations in light intensity which even the adaptable human eye cannot support.

CHAPTER TWELVE

For They Know the Day
and the Hour...

Man in the technological age has created for himself an artificial daily rhythm that has little to do with the conditions of his environment. The fact that we still live according to a 24-hour rhythm is probably due only to the unalterable intervals of our bodily functions. We have long considered ourselves independent of daylight and of changes in temperature. That this belief is mistaken has been demonstrated by the damage to health which undeniably ensues with interchangeable work shifts or with night shifts.

Plants observe natural rhythms very strictly. As we know, this is indispensable to their healthy growth, but it also contributes to the economy of energy and material. Thus luminous marine algae and fungi (see Chapter 15) emit the enzymes essential to their remarkable cold-light radiation only at night. In the daytime they 'switch off' their light, which is economically sound since the intended effect would be lost in daylight.

The time 'agreement' between many flowers and insects is of particular interest. It is at precisely those times when pollinating bees, wasps and other insects are abroad that flowers open and give up their nectar and pollen. As early as 1933 E. Kleber examined this coincidence in detail. Basing the times on averages of measurements taken on various days in July and August, he has shown that both flowers and insects possess internal clocks which function to a large extent independently of the weather, and which quite evidently synchronize. In this way animals avoid visits at the wrong time in their search for food, just as a housewife knows the opening and closing times of her food shops; and for the plants it means that nectar and pollen need be produced only when they will be used.

Such biological clocks go on functioning quite reliably when outward signs of the day's rhythm, such as dawn and sunset or changes in temperature, are artificially suppressed over several days. As though at a word of command the flowers open their blooms at the appointed time, and the insects

82

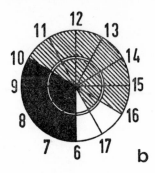

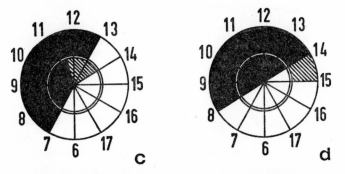

82   Co-operation between plants and pollinating bees is indicated by these clocks (which register only 6 a.m. till 5 p.m.). The black section of the inner circle shows the times of heavy pollen emission by the blossom of four varieties (corn poppy, mullein, vervain and three-coloured bindweed), and the black section of the outer circle shows the time of intensive visiting by bees. Cross-hatched areas indicate light pollen emission and infrequent visiting, and blank areas indicate zero pollen emission and zero visiting.

keep the 'appointment' just as punctually. We do not know how plants and animals developed this capacity for exact time measurement: it was expedient for them, and so they solved the problem by methods which scientists can still not explain except by unsatisfactory hypotheses.

The great Swedish botanist of the eighteenth century, Carl Linnaeus, planted a flower clock based on the different opening times of different blooms. Carefully chosen plants were arranged in segments round a circle in the order of their times of opening. This botanic clock always told the right time throughout the day, for at each hour the flowers of only one species would be open.

'Sleep movements' of many leaves (see Chapter 9), which are designed to prevent an unnecessary heat loss at night, are almost always controlled by the biological clock and not by alternations between light and darkness. Similarly, trees and grasses release their winged seeds and their pollen only at those times of day when they know from experience that the wind will be strongest and air transport most assured. Many fungi and algae release their male and female gametes only at certain times of the day, when both are ready for union. The capacity of chlorophyll to build up glucose is almost completely dormant during the night: it is not needed, so it is 'switched off', exactly as we switch off the radio outside transmission times.

In June 1964 the English journal *Nature* published an article by Dr G. S. Hawkins with the sensational title, 'Stonehenge, a neolithic computer'.

The great ring of Stonehenge, in the county of Wiltshire, about 80 miles southwest of London, was erected by Stone Age men some 3,500 to 4,000 years ago. Slabs of rock up to 25 feet in height and 50 tons in weight were transported from a quarry some 145 miles away expressly for this edifice, which scientists of our day have called 'the greatest riddle of the ancient world'. The riddle has since been solved by astronomers and archaeologists. Stonehenge has been 'deciphered'. That prehistoric place of worship was a sanctuary of the sun and the moon, built in order to locate important positions of the two heavenly bodies by means of precisely sited pillars of rock. It can be proved with certainty that the placing of the stones was not fortuitous, but represented a well-thought-out astronomical orientation apparatus. Stonehenge was an exceedingly accurate sun and moon calendar, which made it possible not only to predict the times of the annual summer and winter solstices, but also to determine the changing position of the moon's rising and setting over many years. By means of this stone calendar, the days between two full moons could be counted; the cycle of 18·6 years, after which the changing course of the moon in the heavens repeats itself, could be followed; and eclipses of the sun and moon could be precisely foretold.

It is not only at Stonehenge that there is a stone calendar thousands of years old. In the northwest of France, in Brittany, mighty accumulations of prehistoric structures are to be found. Stone rows made up of thousands of pieces of rock attract

83

curious tourists and interested scientists every year. These arrangements of stones also had clearly calendrical functions.

Scientists have generally supposed that the orientating lines of all these arrangements of stones coincided only by chance with important astronomical points, for they simply could not credit the builders with the far-reaching astronomical knowledge necessary for a planned alignment. Probability calculations, however, have since proved quite clearly that Stone Age man must have been in a position to erect such calendar structures intentionally. Professor Rolf Müller, the astronomer, who has made these ancient cultural monuments the subject of meticulous study, expresses it thus:

Is it in fact really learning that is revealed here? It is knowledge that was, as a matter of course, the property of an observer close to nature. The spectacle of the heavens taught Stone Age man the movements of the stars in their relationship to one another. He used this knowledge to fix the hour and the calendar, as was so urgently required by a settled agricultural people.

These words, in my opinion, are the key not only to the calendar, but to the whole question of time measurement. Time measurement is necessary for planned action, for adjusting oneself in good time to an expected event. Thus many flowers open not at sunrise, but a little before, because their biological clock tells them that it will soon be day. We human beings of the twentieth century have largely lost the measurement of time in its original sense. When we enter appointments in our diary by date and hour, and ensure the keeping of them by calendars and watches, the dates are nearly always chosen independently of external events. If, for example, an insurance agent arranges an appointment with a customer for the conclusion of an agreement on 27 April at 10.30 a.m., the appointment could just as easily be for 14 September at 3.45 p.m. For that type of business it does not matter. But if the ancient Egyptians wanted to till their fields, the time could not be decided at will. When the annual inundation of the Nile made the soil of the river valley fertile, every preparation for the sowing and planting that were to follow must already have been made. The Egyptians were dependent on a system of time measurement quite inconceivable apart from external events because synchronous with them. Prehistoric man developed this measurement of time because he required it. That the

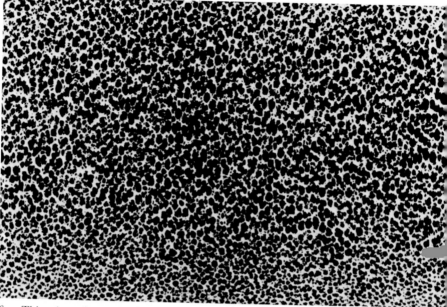

69 This micro-photograph of a thermal insulating board (polystyrol resin foam) shows what the insulation depends on: many small air-filled hollows, which show up black in the picture, connected only by thin, torn bridges of material.

70 A micro-photograph of the spongy texture of an orange skin. The dark pores are not plant cells filled with sap, but air-filled spaces between the cells, constituting an ideal thermal insulation. It is constructed on exactly the same principle as the thermal insulating board, with the same function of protecting against heat and cold.

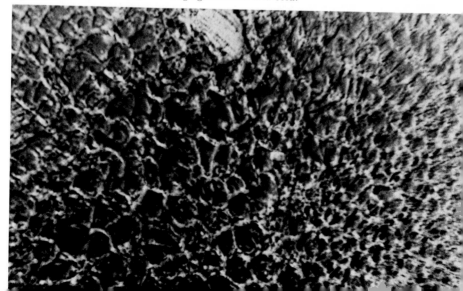

71 A section through the fleshy leaves of the South African *Conophytum* plant shows that they are loosely surrounded by old, shrivelled leaves. This kind of thermal insulation, known to the technologist as 'folio insulation' is commercially applied in the use of aluminium foil for cooking.

72 *Euphorbia obesa.* The plant's spherical shape has proved itself in desert regions. Spheres are the bodies with the smallest surfaces in proportion to content. This reduces evaporation and heat absorption. In addition, spheres receive the light from any and every side: one half is always in shadow. Despite its cactus-like appearance, this plant is in fact a spurge from the deserts of South Africa.

73 Even in the architecture of desert lands we meet the functional sphere: cupolas and the East belong together. This picture shows a marabout's grave in Nefta, southwest Tunisia.

74 Ribbed columns, as a construction for giving shade, have likewise shown themselves to be very effective. The cactus *Neoraimondia gigantea*.

75 A layman would scarcely be able to distinguish the plants in this photograph (the spurge *Euphorbia canariensis*) from those in the above illustration. Yet they belong to entirely different families and are even natives of different continents. Adaptation to similar environmental conditions makes them alike.

76 Where benefit is to be had from ribs or similar providers of shade, as with the cactus crown, a micro-structure often gives real protection against the sun, like the filigree of spines in this micrograph of *Mammillaria herrerae*. The section pictured is about 1 cm wide.

77 a and b The leaves of the wood-sorrel protect themselves by 'sleep movements' from too great loss of heat at night.

78 The mescal cactus withdraws completely into the ground for the winter. The upper part, held fast by the tuber in the ground, is drawn into the earth by shrinkage. Soon the wind blows sand and dust over it: the plant has taken up its winters quarters.

79   Ceramic magnetic rings 1.2 mm in size, threaded on a wire mesh,
made up the core-stores of electronic computers of a few years ago. In the
middle, for comparison, is a complete storage element of a modern computer, about 2.5 mm square.

80 Comparison with a peppermint lozenge illustrates dramatically the minute size of modern magnetic computer cores. With the 4,000 insignificant-looking rings in this picture as many storage spaces can be constructed in an electronic 'memory'.

81 View of a modern computer room. In the foreground to the right is the central store which, with extensions, can take from 500,000 to 2,000,000 0/1 differentiations.

83 Prehistoric man's close-ness to nature may have given him a sensitivity to movements of the sun and moon shared by many plants. Megalithic calendars like this one at Lagatjar, Brittany, are frequently to be found in northwestern Europe.

84 The different structures of two primrose flowers of the same species (*Primula kewensis*) show the mechanism for preventing inbreeding: pollen from the left-hand flower can only make its tube grow through the short style of the one on the right; pollen from the right-hand flower can only achieve the same thing with the long style of the left-hand one. Self-fertilization is thus precluded.

86 Living stones (*Lithops*) of South Africa adapt themselves exactly to their environment in form and colour, and even in texture. In this photograph it is far from simple to find all of them.

87 Matching of form and colour with the environment as a protection against enemies is often a matter of life or death for the soldier too. The picture shows a well-camouflaged mountain infantryman.

88 On the road between Sâo Vicente and Porto do Moniz in the north of Madeira, a dense company of mosses, rosettes of native perennials, and grasses, clothe the perpendicular walls of rock, bare but dripping with moisture.

89 In the dried-up clay and gravel plains of the north and central Sahara, where no rain falls for years on end, the blue-green cushions of *Anabsis aretioides* are often the sole evidence of the toughness of plant life, as here in the desolate plains round the oasis region of Tafilalet. This plant nestles close to the ground, and resembles a close-cropped hedgehog!

90 On the completely smooth surface of a giant saguaro this pineapple plant (*Tillandsia dasyliri-folia*) germinated, grew over the years into a full-sized plant, and is here seen in full flower. It was given no nutriment by the host plant—all its food and water was airborne.

91 The great leaves of this tree-fern (*Platycerium* 'Wilhelminae Reginae') form a 'pot' which traps dust and humus.

92 At the foot of the Torri di Vajolet, east of Bolzano in the Southern Tyrol, at a height of over 8,500 ft, plant communities cover the inhospitable dolomite scree of this alpine desert. A characteristic plant of these scree slopes is the mountain avens (*Dryas octopetala*).

93 In the sterile ash on the black slopes of the volcanoes San Antonio and Teneguia, in the extreme south of the island of La Palma in the Canaries, inconspicuous scree plants flourish, mostly low-growing spurge.

94 In the central highlands of the island of Lanzarote in the Canaries, solidified lava fields cover large areas. Often there is no rain for a whole year, and the rare downpours quickly flow off the smooth rock. In this inhospitable region grows the camel-thorn *Aeonium lanzerottense*, a little rosette plant only to be found here, and a number of lichens.

95 In the shallow lagoons of the Camargue, there are only extremes: inundations or drought, wet mud or cracked, rock-hard earth. In the foreground, the white areas indicate deposits of almost pure cooking salt. To this environment of 'brine-water' and dry, blazing sunlight the samphire (*Salicornia*) has adapted itself.

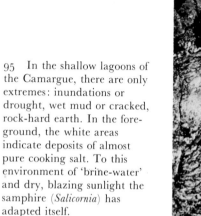

96 On every ledge, however small, of the tall, strangely shaped stalactite columns in the Cueva da Artà, Majorca, a black film indicates colonies of bacteria which draughts of air have wafted into this lightless world.

97 Roots of a sycamore reach straight down like twenty-foot black cables through a cave in the Höllgrotten, Switzerland, seeking floor-level and the earth they need.

98 The age of many dragon trees (*Dracaena draco*) on Tenerife, such as this venerable specimen near Tacaronte, is reckoned to be as much as six thousand years.

99 The long, narrow leaves of the sundew are covered with sticky little drops. Towards the middle of the plant and to the right, near the curled-up leaf-end, can be seen a captured fly.

100a and b   A group of Japanese gill-fungi (*Mycena lux-coeli*), with nothing very unusual about them as long as it is day ... but at night-time they give out a ghostly light, like little Chinese lanterns.

measurement of time is no end in itself, but an indispensable means of adapting oneself to the order of the environment, is a fact that we today frequently overlook. It is still a self-evident fact for the scientists of the space travel laboratories, who can start a moon probe, for example, only at certain times, if they are to link up again with the earth satellite. Biologists, who have long neglected in favour of more obvious spatial considerations the study of the adaptation of man, animal and plant to the rigid temporal order of the environment, seem gradually to have lost their feeling for the significance of time measurement.

Only recently has it become known that plants can carry out long-range time calculations as precisely as the prehistoric astronomers of Stonehenge. Similarly, we have only just discovered that they allow for the movement of the sun and the moon in their internal calendar, and that they actually take measurements of the length of the day.

To return to the two examples of the insurance agent arranging an appointment, and the peasant of the Nile valley: relative time measurement is enough for the insurance agent. If he and his client have synchronized watches he can arrange as the time of meeting '24 days and $7\frac{1}{2}$ hours from now'. The position of the point of reference 'now' is irrelevant. The peasant of the Nile valley, like the astronaut, cannot be satisfied with a relative time measurement. Both need an absolute measurement. Not only must their clock run neither fast nor slow, but it must also be set to local time. Plants face the same necessity: their clock must not only measure relative time intervals, but must also tell them at what time the sun will rise.

To what extent plants fix dates beforehand and plan their behaviour accordingly we can see from a species of brown algae, *Dictyota dichotoma*, which grows in shallow coastal waters. Guided by a precise inner 24-hour clock, it releases its male and female gametes only at certain hours of the day – but not of just any day, for an internal sun calendar restricts it to the late summer. An inner moon calendar also influences the procedure: the gametes are released only twice within a moon-cycle, at intervals of 14 or 15 days. Since all three conditions coincide very rarely, this species of algae liberates its reproductive cells over only a few brief hours in the year. The odds in favour of male and female gametes meeting and uniting are a thousand times greater during this carefully defined brief period than they would be if the cells were released over longer periods. Male and female gametes thus have definite 'appointments' which they observe with painstaking exactitude.

But even that is not the chief advantage of this detailed planning. It also guarantees that the reproduction takes place at low water, and then only at the spring tide, when the water-level is particularly low. Conditions are then most favourable for reproduction. Brown algae thus use a moon calendar because they need it; for the tides are governed by the moon. Laboratory experiments have proved that the plants are not simply using direct observation of the sun and moon, and waiting until the light of the stars indicates the position of the heavenly bodies most favourable for their purposes. Even when the algae were deprived of both sunlight and moonlight they kept to the appointed times. They must therefore possess an inner clock and calendar which enable them to choose the right times even if, for example, the moon is obscured by clouds.

The ability of plants to measure exactly the length of the day is another proof that they do not simply redetermine their rhythm each day according to the alternation of light and darkness over 24 hours. Anyone wishing to ascertain how long it is light and how long dark over a 24-hour period, in order to draw conclusions as to the season, needs a very precise 24-hour clock, and must be able to compare exactly such external time intervals as the duration of light with the course of the inner clock. The incredible precision with which plants have been able to carry out such measurements has been demonstrated by varieties of Javanese rice. Since Java lies very near the Equator, differences in the length of day never vary more than 48 minutes over the whole course of the year. Fractions of this time span, therefore, must suffice the rice plants for the determination of the season. Indeed, it has been proved that a difference in daylight length of one single minute causes a speeding up or a slowing down of more than a whole day in the development of the plant. To detect one minute of light more or less in the course of 24 hours is a precision achievement in technical measurement with a marginal error of 7 parts in 1,000. The long-term calculations of the brown algae, when determining the most favourable hours of the year for emitting gametes, are accurate to within 5 parts in 1,000.

Seasonal clocks or calendars are not rare among plants. Many seeds retain their capacity to germinate for years or even decades, but can only sprout at certain seasons. The seed of the hayrattle (*Rhinanthus alectorolophus*) becomes capable of germination five months after maturity, at the beginning of winter. Its capacity for germination is then maintained for a few months. It ceases in April of the following year, returning five months later for the next winter period. This is of course a wonderful adaptation to the temporal rhythm of the environment: the seed is capable of germination only when the appropriate climatic conditions exist. However, it is not the climatic conditions themselves that regulate germination capacity, but an internal seasonal clock that works independently of them, whatever the conditions and temperatures operating on the seed. Otherwise, the seed could react wrongly to an occasional bout of abnormal weather and perhaps germinate in the middle of a cold and damp summer month, so that its period of fastest growth would coincide with the winter. The seasonal internal clock effectively prevents that. Such hereditary timers often function for periods of more than a year. Ash tree seeds, for example, become capable of germination only eight years after maturity.

We do not know by what technical means plants measure time: the technique was essential for survival and so they learned it, as did Stone Age man.

CHAPTER THIRTEEN

Inbreeding and Incest

Development is the form of progress most favourable to the environment. Constructs, too, can be favourable, but often they are not. Hitherto I have written of constructs in a purely technical sense, indicating when they were related to biological solutions, when intelligent, and when not. But the concept 'construct' can take on a wider meaning, connoting a stiff, theoretical quality, alien to reality; in that wider sense much is constructed, or artificially put in train – even marriages.

The economically minded feudal system of past ages demanded biologically senseless marriages of convenience, 'suitable' to the rank of the parties concerned. The motive was commercial. Man and wife did not come together from natural liking for one another; their union was consciously engineered, 'constructed'. In the course of generations the inevitable result was the intellectual and physical decadence of the nobility, for continuous inbreeding within closely related clans hinders the healthy development of the racial heritage and ends in degeneration.

Degeneration is a biological problem, and biological problems are even less capable of solution through the constructs of abstract reflection than are technical problems. A typical 'construct' solution is the relief of the ills produced by degeneracy. But this leads to dependence on aids, crutches, prostheses and medicine.

A biological solution of the degeneration problem consists entirely in preventing things from getting to that point. As this is something of a digression from the predominantly technological theme of this book, let me indicate, by one example only from the multitude that could be given, the means developed by plants to obviate biological problems.

Inbreeding is as undesirable for many seed plants as for man. Inbreeding, i.e. the fertilization of the female gametes of a flower with the pollen of the same flower, is technically not possible with very many flowering plants. They are able to prevent what is genetically undesirable from happening.

Observation of orchids, for which inbreeding is possible, has clearly shown that seedlings from mutually pollinated blooms are always superior to those from self-pollinated ones, and that self-pollination can show ill effects in a single generation. Clearly, then, preventive measures against inbreeding are genetically effective. Darwin pointed this out a century ago, and the saying, 'Nature tells us in the most emphatic manner that she abhors perpetual self-fertilization,' is still familiar to-day to botanists under the name of Darwin-Knight's law. The ways in which flowering plants prevent self-fertilization differ in details, but all the methods can be reduced to two principles.

If, during pollination, a pollen grain lands on the stigma at the tip of the style a tube grows out of it which works its way down through the style towards an egg-cell in order to fertilize it. According to one principle, this also happens when the pollen lands on the style of its own flower, and this is self-fertilization. But in that case the fertilized egg-cell or, a little later, the young embryo aborts. According to the other principle the pollen tube does not reach an egg-cell of its own flower, because in the style it is chemically or mechanically prevented from growing. Where this is the case, the blooms can often be recognized by their external appearance. Thus primroses and many other plants develop two different kinds of bloom, those with a long style and those with a short one. If pollen from a bloom gets onto its own style, the pollen that develops is not able to grow through the style and press forward into the embryo sac to reach an egg-cell. Successful fertilization is only possible when the pollen of a flower with a short style gets to one with a long style, or vice versa, as is indicated by arrows in the plate. Prevention of inbreeding is therefore complete.

84

CHAPTER FOURTEEN

Adaptation for Survival

Efficient living in a given environment presupposes the best possible adaptation to that environment. Adaptation means specialization, and specialization means division of labour. That is why division of labour leads to efficiency. All the adaptive measures that we meet with in plants are, at bottom, instances of division of labour: the root, the stem, the leaf, take over quite definite functions of importance to the plant, and are for that reason nourished by the other organs. Buttress roots in a swamp, climbing roots, aerial roots, insect-capturing leaves, defensive thorns, urns storing water vapour, glider seeds, these are all specialists within a framework of divided labour. With highly developed forms of adaptation, division of labour can transcend the limitations of a single plant. Different plants, or even plants and animals, can come together in a real association of interests. Orchids are an example.

The nineteenth century was the heyday of orchid hunters. At first the plants were rare in Europe, besides being very delicate, with the result that first-class specimens commanded extremely high prices. So there were always adventurers who, at the risk of their lives and often with inadequate equipment, plunged into tropical forests, felled giant trees with primitive tools, and salvaged the precious plants from the topmost branches. In addition to the usual dangers of such expeditions, the orchid hunters often had to bring their valuable cargo to the coast down unknown rivers in log canoes, at the risk of loss from capsizing or even death by drowning. No wonder that in Europe people tried to grow the plants from seeds. But the seeds refused to germinate. It was not until 1904, almost 200 years after the discovery of the first exotic orchids, that the French botanist Noël Bernard succeeded in solving the puzzle of orchid germination. He was able to show that tropical orchids in their native habitat germinate only with the aid of a fungus which later, when the plants are fully developed, lives in their roots. The minute seed of the orchid has so little food reserve at its disposal that it needs a wet-nurse, as it were, if it is to germinate.

This is where the fungus is helpful, growing out of the orchid root as a fine web of mycorrhizal filaments which penetrate the surrounding humus, there to turn waste products to account. The orchids profit from this gain in food supplies, and for their part provide the fungi with the sugar which they cannot build up for themselves. Orchid and fungus thus form, in the language of the market economy, a partnership or, as the biologists say, a symbiosis. Together, they can adapt themselves better to their habitat high in the branches of the forest trees, often 100 feet or more above the ground, and so survive better.

A very similar but even more interdependent community is formed by a variety of microscopic spherical green algae together with fungi. These two organisms are so closely bound to one another that together they look like a single plant: lichen. The places where lichen grows show how exceptionally advantageous the symbiosis is for both parties. Where neither green algae nor fungus could exist alone, they are able to grow and flourish together: on bare stone, on perpendicular stone walls, on withered tree bark, on pure sand, inside stones, in the climatically most unfavourable regions, the highest parts of mountain ranges, even in the Arctic.

How do the two partners work together? A layer of the spherical green algae is closely woven round by the filaments of the fungus, which can decompose even sterile stones chemically. Thus the algae are provided with important released mineral substances. Besides this they retain the atmospheric humidity. The green algae for their part build up sugar with the help of water, air and light, and provide the fungus with it. This unique co-operation makes both partners largely independent of most environmental factors. It is an extreme form of adaptation to hard living conditions.

Symbiotic contracts are also formed between plants and animals. A number of so-called 'ant-plants' are found in the tropics of both hemispheres – plants so well adapted to a common existence with ants that their actual structure exhibits specially developed peculiarities. One tropical species of acacia (*Acacia sphaerocephala*) bristles with thick brown-black thorns about an inch and a half long. But these powerful-seeming weapons are not so effective as would at first appear. In fact they can be quite easily pressed between finger and thumb, for they are thin-walled and hollow. Nevertheless they have a protective function, for if they are damaged, innumerable ants will emerge and attack the aggressor. Thus they protect the acacias very effectively against consumption by animals. The benefit is mutual: not only does the plant offer the ants a nest safe from floods in its hollow thorns, but also it provides them with sugar, as well as fatty oils and albumen secreted by special organs.

Symbiosis with ants, but in an entirely different style, is also typical of many South American pineapple plants (*Bromelia*). In the flood regions of the Amazon and its tributaries, the water level often rises by several yards. This makes life impossible for ants on the ground, as they would drown every year in their millions. So the ants of those regions have moved into the 'upper storeys' of the forest, secure from flood, and build their nests on trees. They carry earth up, stick it firmly to the trees with their excreta, as with putty, and – expressly in the interests of this unusual nest structure – lay *Bromelia* seeds on it. The plants sprout rapidly, cling to the supporting trees, and lend considerable solidity to the entire edifice. In these gardens, designed and planted by themselves, live the ants. But the *Bromelia* likewise profit from the association, for their gardeners have planted them precisely where they too are secure from floods. Once again it is common adaptation, combined with division of labour, that makes symbiosis a highly advanced form of adaptation to hostile environments.

The desert and steppe regions of South Africa, the veld, are the home of large numbers of plants which practise adaptation to the environment by striving to be as like it as possible.

Rolf Rawé, a well-known botanist, plant collector and authority on South African drought areas, is a veteran of many treks over the veld and has written a highly interesting book on its plant life, with many photographs of these plants in their natural environment. Anyone accustomed to books on garden plants or wild flowers will find the illustrations of Rawé's book remarkably unobtrusive, for in more than half of the plates one has to search for the plants before one can even see them. The cause is neither bad photography nor cheap printing but the plants themselves. In form and colour they are often so like the ground on which they grow that one can scarcely distinguish them from it, even from three feet away.

Considerable regions of the Little Karroo desert and Namaqualand consist of cracked ground thickly sown with white quartz pebbles. Here – and often only here – small fleshy plants

85

85　High above the flood levels of the Amazon, *Bromelia* seeds, carried up by ants, germinate to form a protective 'garden' for both plants and ants.

emerge, all with bodies noticeably white, and frequently as round as the pebbles between which they grow. They belong to the most varied plant families, but are very much alike in their capacity to imitate the natural ground.

In other parts of the Little Karroo, e.g. in the vicinity of Prince Albert, and in the Ceres district, there are rocky patches with very dark quarry stone. Here a quite different kind of plant is to be found. Its body is not white and spherical, like light-coloured gravel, but dark grey and irregularly angular like the granite on which it grows, or wrinkled and shrivelled up like crumbling stone.

Again, small plants with thick white warty leaves, that look deceptively like weather-worn limestone, are found on bare mountain tracts, and in their natural environment can only be detected with great difficulty.

Among the best-known dwarf plants of South Africa are the 'living stones' (*Lithops*). There are at least fifty different varieties that appear only on stony ground, and in their original environment take on to an extraordinary degree the exact shade, and frequently also the surface structure of the pebbles, gravel or rocks among which they grow. They conceal themselves as skilfully and on the same principle, as a soldier in the field. By assuming the same colour and the same structure as the neighbouring rock, both are unrecognizable at first glance.

What is the purpose of this effective camouflage? It has the same meaning for plants as for the soldier: protection from enemies. In desert regions with little vegetation the few plants are forced to make themselves invisible to famished plant-eaters, especially in times of drought. If they were easily visible, it would mean certain extermination within a few years, but their unusual camouflage saves them from being eaten. Only for a short time of the year can the living stones of South Africa be recognized from a distance – when they are in bloom. But during that period – the rainy season when the whole desert is verdant and in bloom – grazing animals are able to find plenty of other food.

And the angel of the Lord appeared unto him in a flame of fire out of the midst of a bush: and he looked, and behold, the bush burned with fire, and the bush was not consumed.

This biblical statement of an early botanical observation, as well known as it is striking, occurs not only in Exodus III, 2, but also is mentioned in Deuteronomy XXXIII, 16, and in the Acts of the Apostles VII, 30.

What is the truth behind this burning and yet unconsumed bush in the desert, which understandably captured the interest of the biblical people? Was it a miracle, or a reality?

That bush existed then, and still exists today. But there is no need to go to Mount Horeb (now called Ras es Safsafeh) in the Sinai mountains to find it. The dittany (*Dictamnus albus*) grows sporadically around the Mediterranean and in Central Europe on dry, calcareous or volcanic soil. It is about three feet high, and is also known as burning bush, gas plant, savory, or caper spurge. The dittany is just such a burning bush as we read of in the Bible. On hot windless days it discharges a highly volatile oil which can easily be set alight. This oil burns with a blue flame, without the bush being damaged in any way. In intense heat there may even be spontaneous ignition.

The biological significance of the oil exhalation (for which the plant possesses special glands) has not been exactly established, but we may conjecture that it is a case of 'fighting fire with fire'. Strangely enough, the emission of light inflammable oil effectively protects herbaceous plants against being burnt up in hot regions. First, heat evaporation through the emission of oil lowers the temperature in the leaves; secondly, spontaneous combustion in great heat, where it ignites a carpet of dry grass, protects the plant against flames from fire in the vicinity, just as forest fires can sometimes be checked by a burnt-off strip.

If you find it hard to believe that succulent plants, secreting light combustible oils, are themselves protected from damage by fire when the oil-charged air around them takes fire, try the following simple experiment. Steep a well-damped linen cloth in methylated spirit, hold it in a long pair of tongs or scissors, and set it alight. The highly volatile spirit (corresponding to the volatile oil of the plant) will burn quickly with a blue flame, but the damp surface of the cloth (corresponding to the succulent body of the plant) is so much cooled through the evaporation of water in the heat that the flame cannot spread to it, and goes out as soon as the methylated spirit is consumed. The cloth remains intact.

Sand, sand, nothing but sand as far as the eye can see – such is the picture conjured up by the word 'desert'. There are deserts on every continent but most of them do not at all correspond to this conventional picture. Even the archetype of sandy deserts, the Sahara, is far from being a single field of dunes. Vast rocky mountain chains alternate with wide, flat, pot-holed clay fields, monotonous gravelly plains, or salt pans scintillating in the heat. Grandiose landscapes of canyons suddenly give way to extensive table mountains. Strange, weather-worn towers of rock stand erect over plains where dust-storms may darken the sky for days on end. Then their place is taken once again by fields of gravel, covered to the horizon with red, black or blinding white pebbles, like a gigantic almond cake. The deserts of the picture-books with their wandering, singing sand-dunes also exist, of course, but these are only the deserts of tropical lands. The crisp and frosty waste lands of the poles, the inhospitable high mountain regions, volcanoes and subterranean caves are desert regions too.

Plants which have won the deserts of the earth for their living space have received from botanists the honourable title of 'pioneer plants', not without reason, for their ability to deal with some of the most difficult and hostile conditions on earth is extraordinary. Singeing heat, which would wither and burn up unadapted plants in a few hours; scourging sand-storms which erode rocks like gigantic sand-blasters; droughts that last for years; temperatures which fluctuate more than 50° or 60°C in twenty-four hours; temperatures which remain at 50°C below 88 zero, and permanent fields of snow; hot volcanic springs of near-boiling dilute sulphuric acid; barren, vertical rock-faces without a trace of humus; heaps of scree several yards high from which every shower of rain immediately runs off, or which move slowly down mountain slopes, crushing everything in their path; marshes with water salty enough to pickle herring; or darkness complete and absolute: those are the problems with which pioneer plants have to cope, the menaces that they have overcome by their extreme flexibility.

89   In the driest areas of the Sahara, there grows the cushion plant *Anabsis aretioides*, which owes its survival in the extreme conditions of its habitat to its sponge-like shape. The temperature of the stony ground around the cushions may easily rise at midday to 70°C, which would quickly kill the root crown of most plants; among the shady balls of the *Anabsis*, however, it is relatively cool. Besides, only the surface of the cushions is alive. Within, the dead parts of the plants form a kind of humus ball

which stores the scanty moisture and ensures that the substances built up by the plant can, after decaying, be used again. *Anabsis*, therefore, in course of time forms its own 'flower-pot' with its own organic potting compost in an inorganic environment. It grows over it, and its hedgehog form protects it at the same time from the abrasive effect of the dust-storms.

Desert plants with 'flower-pots' of a quite different kind occur in tropical America, where they survive in still more extreme conditions than the blue-green hedgehogs of the Sahara. These ingenious members of the pineapple family (*Tillandsia dasyliri-* 90 *folia*) can sometimes grow where there are not even stones and clay, for instance, on the side of a saguaro cactus. Occasional rain, mist and dew bring the necessary water, and the leaves, arching downwards to form a 'flower-pot', pick up dust, which brings the necessary mineral substances. This aerial way of life is not confined to pineapple plants. As we have seen, many orchids, especially the big tropical tree-ferns, act in a similar way.

There is no lack of rainfall in the stony deserts of the high 91 mountains, of course, but water quickly flows off the steep rock faces or the loose screes of the moraines, so that in order to survive plants must be adapted to irregular supplies of moisture. Like the dwellers in tropical deserts they have to contend with slashing storms, and like the plants of hot, dry regions they have only a few weeks in the year for growth, flowering, development of fruit, and dissemination of seed. Tropical desert and dry-region plants remain dormant for most of the year; while mountain plants generally sleep for just as long under a cover of frost and snow. There is no spring and no autumn in the regions of eternal snow, and summer lasts only for a few weeks, if it does not fail completely. In that short time the glacier crow-foot, the highest-flowering plant in the Alps (sometimes more than 11,000 feet above sea-level) completes its entire annual life-cycle.

In recent years botanists of the University of Innsbruck in Austria have investigated the living conditions of high alpine flora on the Nebelkogel at Sölden, in the Tyrol, 10,470 feet above sea-level. In 1964 the summer was very good: 68 productive days for the plant. In 1965, which was cold and damp, there were only 31 days. 1966 was just as unfavourable: in that year many plants remained completely buried under the carpet of snow. Only in 1967 did the weather relent; after 33 months of almost uninterrupted life in the dark under the snow the dwellers at high altitudes got a chance to recover. Many had completely retrenched all their flower buds and, as it were, ceased to breathe in order to survive. Others that had not been covered by the snow during severe cold spells had lost leaves; and the loss of a single leaf must be considered in its relation to the production loss in the short period of growth during a whole year. Even in 1967 the warm days were repeatedly interrupted by waves of cold and fresh heavy falls of snow. But the blossoms, though checked in their growth by frost and glare, appeared immediately after the first melting of the snows and, exactly 30 days after the blossoming, the first fruits were ripe.

Soil conditions in Alpine deserts are as hostile as climatic conditions. But plants have learned to adapt themselves to these also. They grow out of crevices in the rocks and on steep slopes over which a compact covering of snow crawls slowly down every year, not more than 12 inches a month, but with enough force to bend iron fence posts two inches thick as though they were twigs. The plants grow even in the loose scree which, like the snow, creeps down the slope, burying and crushing everything under it. They manage to pass through the downward-moving scree, to crawl over it, to stretch themselves through its interstices, to cover it over or even to block its movement, and in one way or another to adapt themselves to the restless, grudging soil.

In mountainous regions where environmental conditions eventually prove too severe for flowering plants to live, lichens and mosses are quite at their ease, flourishing on Himalayan summits 23,000 feet above sea-level. The heat of the sun matters as little to them as an extreme frost. Their habitat, apparently, may be anywhere. Many can grow even at 24°C below zero – very slowly, it is true, not more than a millimetre a year, but constantly. If it becomes altogether too cold, the lichens become dormant. At this point the temperature may quite well be 200°C below zero, as laboratory experiments have shown. It does not affect lichens. They are capable of using very short periods of warmth, for in a few minutes they can switch from the absolute rest of freezing conditions to full activity. As they are also extremely modest in their demands, and can live even on sterile quartz, they are quite at home in Arctic regions, and often provide the chief foodstuff of the Lapland reindeer breeders and their herds in the cold season of the year.

Are there more extreme conditions than those of the central Sahara, the highest summits of the Himalayas, or the poles? One can scarcely believe it, but there are. In the immediate vicinity of volcanoes life becomes a hell. Hot sulphurous fumes stream out of the ground; boiling acids seethe in craters; bubbles rise in pools of mud, burst, and release poisonous gases. A few inches under the earth's surface the ground is so hot that one can fry eggs on it or set brushwood alight. At the same time, atmospheric humidity is at a maximum. Often a glowing rain of hot ashes will descend as dust or as a shower of jagged, inch-thick bullets. Fiery streams of lava flow past. Here the limits for even the toughest plant life are finally reached. Plant saps would boil at temperatures above 100°C; every plant would necessarily suffocate in a constant sulphurous atmosphere. Yet on the edge of this inferno life has settled, and adapts itself to the encroachments of hell. Once again it is lichens and algae that deal with the near-impossible. Blue algae can survive in hot springs at a temperature of 80°C, pebble algae flourish luxuriantly at 50–60°C, and exist even at a constant 94°C, and so can live permanently in nearly boiling water. Thermal springs in Java have blue algae colonies established in hot water at a temperature of over 60°C, with in addition a high percentage of sulphur.

But higher plants, too, have accustomed themselves to living in these conditions. Some species of heather (*Ericaceae*) and relatives of the bilberry even venture into the vicinity of bubbling mud masses. Thanks to a unique adaptation, their roots can tolerate a temperature of up to 75°C, for their whole surface is covered by a thick layer of cork which contains a good deal of air and works as an efficient heat insulator. Tree heather copes in another way with the high temperature prevailing a little under the surface of the ground. This plant, which is taller than a man, has roots that grow only a few centimetres into the ground, just far enough for the heat to rise above 23°C. At precisely this temperature threshold the roots bend sideways and crawl away under the surface of the earth. They have developed the ability to take very exact temperature measurements, about equal to those of a good thermostat, in order to avoid overheated regions.

As regards quality of ground, volcanic flora must be hardy for the earth is very acid. The most extreme values have been measured around the roots of certain marsh reed plants which manage to grow on the Tecapa volcano in El Salvador, in mud containing sulphuric acid. The ground moisture there is more acid than lemon juice or vinegar, and is described by chemists as a strong acid (pH 1·6). Volcanic ground is not exactly poor in salts either. In many areas, plants must deal with a high content of alum. This they manage because they have learned how to store up considerable quantities of aluminium, which is contained in alum. It is an ability that makes them quite unique, for all other plants are able to process only very small quantities of aluminium.

Life is not simple on the border of craters, even when not immediately endangered by volcanic gases and heat. Volcanic ash, often at a fairly steep angle on the slope of a crater, is by no means ideal soil, nor are crumpled, folded crusts of lava. Yet even here modest plant communities are always to be found, with thick-leaved plants, camel's thorn and lichens among them.

93, 94

It is not only new volcanic ground that is marked by high concentrations of salts. Seashores are continually washed by salt water, which evaporates in great quantities in shallow pools, often leaving white crusts of salt in the dry season, or on cracked and parched ground, as in the Camargue in the South of France. The goosefoot plants which maintain themselves here must tolerate not only floods lasting months, followed by absolute droughts under a blazing sun, and ground hard as rock, but also quantities of salt into the bargain. This is as excruciating as if one were to water indoor plants not with rainwater but with pickle-brine.

95

However difficult living conditions may be in tropical deserts, high mountains, polar regions, volcanic areas or salt-pans, one thing never fails the plants there: light. Without light there can be no life for green plants, and this explains their complete absence from caves. It is often so dark at cave entrances that only a few species can maintain themselves there – those that have no great need for light. Stinging nettles are often to be found at such places, because in the long run they can manage with only 1 per cent of an average day's light. Farther in, the spleenworts (*Asplenium trichomanes*) flourish, as well as a few mosses and various algae. A two-thousandth part of daylight just suffices to keep them alive. (Species of algae are able to live in the semi-obscurity of the oceans, even at a depth of 300 feet, but they, too, fade out in absolute darkness.)

Only fungi and bacteria can do more. Having travelled underground in flowing water or currents of air they manage to feed on organic substances that have either gone by the same route or been carried there by cave explorers. The exceptionally delicate, colourless filigree of the mould fungus grows on all plant and animal refuse in the damp and windless climate of caves. The snow-white cotton-wool balls of the mucor fungi can grow as big as footballs, and are so sensitive that at the slightest touch, or even at a flash of light, they fall to pieces.

Bacteria are always present in a cave if it has the merest breath of air. They can be observed particularly well in the warm grottoes of Mediterranean countries. In still air they take days or even weeks to fall to the ground, or onto a ledge or a stalactite, where they form a black deposit on the upper side.   96 Overhanging parts are always free of them.

Twice during my visits to almost fifty different caves I have met with a botanical curiosity which deserves mention in a book about plant adaptations, even if it is not really typical, or specifically connected with plant life in caves. This concerns penetration by tree roots (in both cases sycamores') into caves from the earth above. One example was in the Höllgrotten at   97 Baar, Switzerland – well-known for their bewilderingly varied stalactite forms – and the other was in the cave of Pech-Merle at Cabrerets in the *Département* of Lot, France, where many traces of prehistoric man have been found. The darkness and the high degree of atmospheric humidity represented ideal life conditions for the roots. Yet the new environment was unusual: there was no earth. The roots at once responded to the new situation. They did not divide as usual but grew (in the Höllgrotten across a space of 20 feet) with uniform thickness directly towards the floor of the cave in order to reach the earth where they belonged.

CHAPTER FIFTEEN

Plant Freaks and Oddities

To watch the grass growing is not just a figure of speech but an actual possibility. Pluck an ear of rye on a warm summer's day, and you can see with your own eyes the stamens pushing themselves out of the ear at a speed of nearly 2 mm a minute. Tropical bamboo shoots come out of the ground a little more slowly, but achieve a far greater quantity of production. Shoots several centimetres thick lengthen themselves by 2 mm every five minutes, or up to 4 inches in only 4 hours.

But the record for growth is held by the fruit of the plant which in its native Brazil is poetically called 'the lady of the white veil' – a tropical fungus with the botanical name of *Dictyophora*.

Its whitish egg-shaped form grows very rapidly, then tapers to a point and bursts; from this point a stalk with a little green hat thrusts out at the rate of 5 mm a minute. Every now and then, during this rapid growth, there is a crackling sound of tearing stalk tissue: one can literally hear 'the lady of the white veil' growing. Soon, at the rate characteristic of this remarkable little plant, a filigree of fine white mesh grows out of the little hat at the end of the stalk: this is the 'veil'. Unfortunately, fair though the 'lady' may be, she has a very unladylike odour of putrefying meat. The smell attracts flies, and the flies disperse the spores.

This fungus passes away as quickly as it has come. Shooting up in record time to 4 inches, it lives for a single night and by the morning has collapsed to a little heap of filth.

More lasting are a series of quick-growing tropical trees. The big leguminous tree *Amherstia nobilis* puts forth brushes of leaves up to a metre long within a few days. Eucalyptuses can grow to nearly 50 feet in three years, and *Albizzia moluccana* of the Asiatic tropics grows so rapidly that plants only one year old measure from 15 to 20 feet. Six-year-old specimens can be as much as 80 feet high, and their stems at the height of a man are 8 to 10 inches thick.

Not only do eucalyptuses grow very quickly; they also grow to a tremendous height. The giant of the Latrobe river in Tasmania has a height of 550 feet, 21 feet taller than the highest church tower in the world, that of Ulm Cathedral.

Fig trees develop the mightiest tops. The monarch of the tribe is in the Botanic Gardens of Calcutta. Although its height is only 85 feet, it is 65 feet thick. The crown has a circumference of 975 feet, this enormous dome being held by 562 aerial roots. This tree shades an area of 8,400 square yards.

Trees are not the only plants that can develop gigantic dimensions. In Pennsylvania a bilberry plant at least 1,200 years old was discovered in 1918; it covered an area of 90,000 square yards. This was no colony, but a single plant. Two years later a still more gigantic bilberry bush was discovered, in the same district, which had overgrown an area a mile and one-third across. It was spreading by subterranean runners at an average rate of 6 inches a year. If growth was always at the same rapid pace, it must have been 13,000 years old, and therefore have germinated in the Early Stone Age. Trees can also live for several thousand years, baobabs, dragon trees, and Californian sequoias being the most longeval.

The vine is one of the plants which in the course of time may grow to gigantic proportions. The biggest grows in Scotland. Planted in 1831, it has extended itself over an area of more than 4,700 square yards. The most fruitful vine is probably one near Graz in Austria – one single plant which gave 44 gallons of wine in 1935.

Among the biggest blooms in the world is the *Rafflesia*, a native of the jungles of Sumatra in the East Indies, with a fetid carrion smell. It has a diameter of 18 to 36 inches, while the plant itself develops neither roots, shoots, nor even leaves. Putting out thin threads reminiscent of the meshes of fungi, it lives as a parasite within the branches and roots of forest trees.

Water plants, too, can attain impressive dimensions: seaweeds of the genus *Laminariales* can grow up to a hundred yards long.

Some seeds retain the capacity for germination over remarkably long periods of time. During the air-raids on London in September 1940, the Natural History Museum in South Kensington suffered damage by fire. Some seeds of *Albizzia julibrissin*, the silk-tree, were affected by water from the fire hoses, and in due course germinated. Records show that those seeds were gathered in 1793!

The longest rest period on record for seeds is reported of the Indian lotus: the New York Botanic Garden announced in 1945 the germination of some lotus seeds that had lain in a herbarium for 250 years. Suspended animation for this length of time is an astonishing achievement, and one wonders how the living cell material, built up out of highly complicated combinations of albumen, organic acids and other substances, can last for such long periods unimpaired.

Indeed, many plant seeds are tremendously tough. Dormant air-dried seeds can in many cases be chilled to $-250°C$ without losing their ability to germinate. Lucerne seeds can tolerate heating for hours up to $100°C$ without injury. Red clover seeds can stay for years in pure alcohol, and after that will still germinate.

Bacteria spores live far longer than any seeds of flowering plants: some Egyptian spores which came to light at the excavation of papyrus rolls 3,000 years old proved to be still quite viable.

What is the absolute age record for living organisms? A Canadian biologist found germ cells of unicellular plants in rock from the Early Cambrian period. For nearly 600 million years they had been cut off from air and could receive no nourishment. Even so, they had been able to survive through this tremendous passage of time.

Curiosities among plants include the various insect-eaters with their diabolically ingenious trapping equipment. Of these, the Venus fly-trap has already been described in chapter 8. Others include the sundew, *Drosophyllum lusitanicum*, widespread in Portugal and Morocco, which has a simpler mechanism. Its leaves are thickly covered with glandular hairs, at the end of which are little dew-drops whose delicate honey-like smell attracts insects. If these insects settle, they remain stuck to the treacherous drops as to a glue fly-paper (in fact, the sundew is used in some countries for that very purpose). This triggers the secretion by the plant of digestive juices, which dissolve the victim. The pitcher-plants of the tropical forests of the Indian Ocean islands capture insects with specialized leaves shaped like jugs, which contain digestive juices; some of these are a yard and more deep. Once an insect, attracted by the sweet scent of the trap-jug, has entered the leaf, it is irretrievably lost. It can find no foothold on the smooth walls, for juice continually rolls off them. Flies can settle on overhanging panes of glass, but their adhesive feet fail them on damp surfaces. So they fall to the bottom of the jug, drown in the digestive juice and are dissolved.

Highly interesting flesh-eaters, but little-known except to botanists, are the 250 species of *Utricularia*, water plants that capture small animals living in the water. These simply swallow their prey with little pouches, 0·3 to 5 mm in size, which close with a door, kept absolutely watertight by means of a sill. In a way not yet fully explained, but presumably electro-chemically, the plant pumps half of the water out of these pouches, causing a strong negative pressure inside them. If a water animal, such as a mosquito larva, then touches one of the lever-like bristles that are set like latches on the trap-door, it springs open inwards and the victim is sucked in by the negative pressure. The botanist F. E. Lloyd, who has filmed this operation, found that the opening takes the hundred and sixtieth part of a second. In one fortieth of a second the door closes again, springing back elastically as the balance of pressure reasserts itself. The whole operation works as quickly as the shutter of a camera.

This plant carnivore is quite 'humane', however. Before it digests its prey, which will take anything from 12 to 48 hours, it kills it chemically. The importance of animals for the development of *Utricularia* was first recognized by the botanist M. Büsgen, who established at the turn of the century that plants thus fed grew twice as quickly as those from which animal food was withheld. It has even been proposed to control mosquito larvae in malaria-infested districts by widespread planting of *Utricularia* species.

One of the foremost authorities on carnivorous plants, Strehli, writes:

In digestive activity, insect-eating plants can challenge any animal stomach. Not only the living muscle tissue of insects is digested, but raw, minced or roasted beef or veal. Even strong cheese, tough gristle, nitrogen-rich plant seeds, pollen, fragments of bone, and tooth enamel cannot resist their powers of digestion. Only farinaceous, sweet and sour substances are not digested.

Certain bacteria that live in sea-water, as well as a number of marine dinoflagellates and a few dozen fungi, possess the unusual capacity of shining at night so brightly that one can read by their light.

Swarms of the tiny dinoflagellate *Pyrodinium* live in a little bay near Parguera in Puerto Rico. On new moon nights or under a clouded sky they shine so intensely that the bay is called Bahía Fosforescente, 'Phosphorescent Bay'. They shine particularly brightly when the water is disturbed. A boat's wake, or even the path of swimming fish casts a magic glow like the tail of a comet.

100   A luminous fungus grows near Nakano-go on the Japanese island of Hachijo and in the dark its light can be seen from fifty yards away. Luminescent fungi are also to be found in European and North American woods, e.g. the deliciously flavoured honey mushroom.

Science has not yet been able to explain the importance of luminescence for plants. What has been proved, however, is that their sources of light work more economically than any human lamps, with an efficiency of almost 100 per cent – unequalled even by modern fluorescent lighting, which also diffuses cold light and gives off no useless heat. But it is not only in relation to efficiency that botanic lamps are economical in their working: they are so devised that the light is turned on only when it is really needed, i.e. in the dark.

CHAPTER SIXTEEN

Where Do We Go from Here?

In 40,000 years of development, man has brought technology to its present level, at first slowly, then with increasing speed. Our technological knowledge is now doubling every five years. In those 40,000 years man has created much that makes life easier, and much that makes it more difficult. There is no reason to belittle the many gratifications that technology offers, nor to overlook the price that it demands.

This book has shown that plants, too, solve technical problems by technical means. But they can do this without any problems of noise or garbage, without polluting the air or setting up stress situations that need psychiatric treatment. Yet plant and human technology are often amazingly similar. Where, then, is the difference, and where did we go wrong? What can we do to solve problems by technological means without at the same time creating greater problems? Answers to these questions can often be read between the lines of this book. Here, in a last chapter, they will be summarized.

First let us take stock of the differences between human and plant technology. We shall then see what we *can* do, indeed what we *must* do at once, if we are to live with our environment and not against it.

1   The technology of primitive peoples is necessarily still very close to nature. It meets all the direct demands of the environment. Human societies at the Stone Age level of civilization are therefore understandably the only ones that can maintain themselves for millennia with absolutely no change, and to some extent still continue to do so today. They know nothing about any rise, zenith, or decline in their culture. But since the first advanced civilizations arose, technology has freed itself more and more from the constraints of nature, and gone its own way. The independent 'construct' has emerged from development – as a result, admittedly, of that process of selection which is also nature's way.

2   Feedback to the environment broke down when the con-

struct appeared. Development is only possible if we continually select those principles that are favourable to the environment, and thus adapt to the environment. Constructs go their own way, ignoring all environmental factors which are not immediately relevant to their aim; they ignore the effect of their actions on the environment. Development sticks close to practice, but constructs, because of the absence of feedback from the environment, are emphatically theoretical.

3 Because we are increasingly preoccupied with theory, feedback is often completely lost, and we became unable to act. For instance, instead of eliminating, or better still preventing, pollution of the environment, we waste our time analysing it at great technical and financial cost.

4 Absence of feedback leads to rigidity in the solution of problems. Once a solution is found, it is not continuously adapted to the constantly changing environmental conditions. On the contrary, for a while an attempt is made to adapt the environment to the construct, which inevitably damages the environment, because the natural state of equilibrium is bound to be disturbed. Only when the environment cannot be changed beyond a certain point is the solution of the problem changed, but even then in a spasmodic rather than continuous fashion. For man, who always suffers an equivalent damage and disturbance, the inevitable consequences are stress situations and social tensions.

5 Lacking feedback in technology, man's general understanding of his environment diminishes. Thus scientists are baffled when they see how accurately Stone Age and Bronze Age man measured time and mastered the calendar. This only means that our forefathers observed the environment, as they had to. Lack of consciousness of the environment finally leads to technological solutions that are neither sympathetic to it nor creative, but defensive and destructive.

6 Development is a long-term process. Nature proves this, for natural species do not adapt themselves overnight to a new environment. (Which is probably one reason why many modern biologists have actually excluded the word 'adaptation' from their vocabulary.) That long periods of time are required for developmental processes is an inevitable result of continual feedback, whereby the all-round suitability to the environment

of the product of development is constantly re-assessed. The construct, on the contrary, is a short-term process, and so in the end its product will be faultier and more susceptible to interference than the product of development.

7 Technology, working predominantly with constructs, can change things radically in a short time, while the environment, working with developmental methods, cannot. This means increasingly that technology appears to forge triumphantly ahead while natural adaptation lags behind, faint yet pursuing. This, however, does not represent an advantage for technology: in the last resort, it will have to take into account the natural factors of the environment. So one day it will have to wait till nature catches up with it. Technology will find the waiting painful, for it survives by racing ahead. But if it does not wait it will bring about its own self-destruction, or the total destruction of the environment.

8 As long ago as the early 1960s, the far-sighted author and thinker K. W. Marek, in his book *Yestermorrow: notes on man's progress* (London, 1961), put forward the idea that not only does the machine belong to the environment of man, but man belongs to the environment of the machine. As the machine is a technologically constructed mechanism, while man on the contrary is a biologically developing one, man cannot adapt himself in the short term to the machine. The constructed machine for its part pays no regard to its environment, man. Stress situations therefore are constantly on the increase.

Taking the eight points together we see that all the drawbacks of the human technological system as compared with plant technology are a consequence of man's thinking in constructs. Non-thinking nature works far more practically and intelligently in the long run. She knows no logic, therefore she can never be illogical. But if, from this reflection, we are to conclude that human aspiration is inevitably the cause of his error, the attitude would be fatal to our future. That we can think, we have sufficiently proved. Now we must learn to *re*think things from their foundations. Our next 'constructive' aim must be pliancy and adaptation, conscious pliancy, not fatalism. How we are to proceed in detail may be deduced from the same eight points.

1 Technology must find a way back from the self-glorification

of its construct principles and their disregard of the environment, to the consciousness of man's total dependence on the environment, and thus to the system of development.

2   We must reverse, as quickly and as thoroughly as possible, those effects of technology on the environment that have been harmful, and prevent any future ones.

3   Theory and analysis are not in themselves the cure. The only thing that can help is the action of responsible (and, if necessary, authoritarian) governments. It is not enough to determine the exact dimensions of the various injuries to the environment: they must be removed in the most direct way possible in each case. For instance, measurements of exhaust gas concentration in big towns cannot take the place of prohibiting motor traffic in those towns. Here, too much is better than too little.

4   All solutions of technical problems, especially those that are considered to have proved themselves, must be continually re-examined as to their fitness in relation to the changing environment. The reverse also applies: the demands made by environmental solutions must be kept under constant review. Town planning, for example, is good, and therefore advisable, only so long as the accumulation of dwelling houses and places of work does not overwhelm the resources of traffic control, and does not lead to physical and psychological harm to the inhabitants. In many overcrowded cities these conditions have not been fulfilled for a long time. A government which exclusively furthers the interests of towns is acting irresponsibly, for as long as towns are assisted they will continue to grow. If our new doctrine of development is to be applied, an 'anti-town' law is called for.

5   An understanding of the environment – not the artificial but the natural environment – should be aroused and promoted in every possible way. This means rethinking all the priorities of our daily life from beginning to end. To quote the German Weltbund zum Schutze des Lebens (World League for the Defence of Life):

The politicians and economists of every country will have to recognize that, considering the frightful immediacy of the dangers that threaten us, all so-called 'great' events of today's world (labour conflicts and prize fights, revolutionary troubles and war upsets, monetary crises) and all plans and projects zealously programmed for a 'future' (EEC, space travel, economic and cultural schemes) appear absolutely trivial and ridiculous, and are fundamentally senseless and remote from reality. A politician voting to squander billions in public money on schemes that have nothing to do with the saving of life, is acting irresponsibly and betraying humanity. We must realize the breakneck speed at which we are approaching the point of no return. The hitherto accepted habits and principles of political and economic life now belong to the past. From now on there can be only one common policy for all the peoples of the earth: the saving of life.

6, 7 Technology's golden calf was and is blind progress. Deliberation, and the courage to call a halt to existing tendencies, must immediately be put into practice if our awakening is not to be a traumatic one. Already the electricity supply companies prophesy electricity cuts in the next decade, because of lack of power. This need not have happened, but for the continual production of new electrical gadgets. Drastic limitation of advertising and, if necessary, even prescriptive limitation of production, are the sort of harsh measures which alone are worth discussing, and which must be put through if the worst is to be avoided. Further extensions to the existing power stations are no long-term solution. A continual drain on power supply will lead in the end to the stripping of our reserves.

8 A ship comes to a halt only if it is switched from 'full ahead' to 'full astern', and even when this radical measure is applied, an ocean liner at full speed still takes as much as twenty sea miles to come to a standstill. Slowing or stopping the engines would be as good as useless. Similarly, those people who will be in charge of the task of reining in our runaway technology will have to start with the same idea – and that means all of us. Man long ago reached the limit of his speed in adaptation. Every further technological advance – in the traditional sense – threatens his existence. It is absurd to discuss the possibility of helping him to adapt himself to his technology with the aid of psychiatry, drugs or recreational activities. Such adaptation cannot succeed, as long as technology advances more rapidly than man. And why should man adapt himself to technology? Is not the first task of technology to help man to adapt himself to his natural environment?

To begin to adapt technology to man is but the first step in this direction.

Bibliography and Sources of Information

BACKEBERG, CURT *Wunderwelt Kakteen*. Jena, 1961

BORISSAVLIEVITCH, M. *The Golden Number*. London, 1958

BÜNNING, E. *Die Physiologische Uhr*. Berlin, 1958

CRAIG, R. T. *The Mammillaria Handbook*. Pasadena, 1945

DARWIN, CHARLES *On the Origin of Species by means of Natural Selection*. London, 1859

EBERLE, G. *Gesteinsflur*. Frankfurt, 1963

ETRICH, IGO *Die Taube – Memoiren des Flugpioniers Dr.-Ing.h.c. Igo Etrich*. Vienna, 1915

GATES, D. M., R. ALDERFER AND E. TAYLOR 'Leaf Temperatures of Desert Plants' in *Proceedings of the American Association for the Advancement of Science*, 1968

GENTRY, J. 'Protection of the Environment is Practical' in *Plain Truth*, Pasadena, California, January 1973

GOUWS, J. B., AND J. AALBERS 'Annular Curves of the Osmotic Pressure of Certain Plants on the Cape Flats' in *Journal of South African Botany*, Vol. 35, Part 2 (1969)

GUTTENBERG, HERMANN VON *Bewegungsgewebe und Perzeptionsorgane*. Berlin and Stuttgart, 1971

HABERLANDT, G. *Eine botanische Tropenreise*. Leipzig, 1893

HAMBIDGE, JAY *Practical Applications of Dynamic Symmetry*. New Haven, Conn., 1932

HERTEL, HEINRICH *Structure – Form – Movement*. New York, 1966

HOBHOUSE, L. T. *Development and Purpose*. 2nd ed., London, 1927

JACOBSEN, H. *Handbuch der sukkulenten Pflanzen*, Vols. I–III. Jena, 1954

KÜHN, H. *Vorgeschichte der Menschheit*, Vol. II, Cologne, 1963

LINDEMANN, G., AND H. BOEKHOFF *Lexikon der Kunststile*, Vol. II, Hamburg, 1970

MAREK, K. W. *Yestermorrow's Notes on Man's Progress*, London, 1961

MINDT, H. R. 'Die Mathematik der Spiralzeilen und das Gesetz optimaler Ästhetik', pts. 1 and 2, in *Kakteen und andere Sukkulenten*, Vols. 9 and 10. Stuttgart, 1967

MOHR, H. *Pflanzenphysiologie*. 2nd ed., Berlin, 1971

PIETSCH, A. *Unkrautsamen und Unkrautfrüchte*, Stuttgart, 1937

RAUH, WERNER *Die grossartige Welt der Sukkulenten*. Hamburg and Berlin, 1967

RAWÉ, ROLF *Succulents in the Veld*. Cape Town, 1968

REISIGL, H. 'Die Pflanzenwelt der Alpen' in *Die Welt der Alpen*. Innsbruck, 1970

RICHTER, W. *Zimmerpflanzen von heute und morgen: Bromeliaceen*. Neudamm, 1965

SCHIMPER, A. F. W. AND F. K. VON FABER *Pflanzengeographie auf physiologischer Grundlage*. 3rd ed., Jena, 1935

SMOLIK, H. W. *Das überlistete Tier*, Lux-Lesebogen No. 47, Murnau, 1950

SWEENEY, B. M. *Biological Clocks in Plants*. New Haven, Conn., 1962

VARESCHI, V. *Geschichtslose Ufer*. Munich, 1959

Verein 'Hütte' *Des Ingenieurs Taschenbuch*, Vol. I. 28th ed., Berlin, 1955

VOGEL, R. 'Zur Kenntnis des feineren Baues der Geruchsorgane bei Wespen und Bienen' in *Zeitschrift für wissenschaftliche Zoologie*, Vol. 120 (1923)

WACHSMANN, KONRAD *The Turning-point of Building, Structure and Design*. New York, 1961

WALTER L. H. *Die Vegetation der Erde*, Vols. 1, 2. Jena, 2nd edition, 1964

WILLIS, J. C. *A Dictionary of the Flowering Plants and Ferns*. Cambridge, 1966

WHITE, A., AND B. L. SLOANE *The Stapeliaceae*, Vols. I-III. Pasadena, Cal., 1937

ZAHL, PAUL A. 'Nature's Night Lights: Probing the Secrets of Bioluminescence' in *Journal of the National Geographic Society*, Vol. CXL Washington, 1971

Technical literature, documents and personal communications from:

Barnes Engineering Comp., Stamford, Connecticut
Bayer AG, Krefeld, West Germany
Botanical garden and museum, Berlin-Dahlem
Chemische Fabrik Klaus-W. Voss, Uetersen, West Germany
Chemische Werke Huls AG, Marl, West Germany
Dow Chemical, Frankfurt, West Germany
Hexcel Honeycomb, Dublin, California
Hexcel S.A., Welkenraedt, Belgium
IBM Deutschland GmbH, Stuttgart, West Germany
Landesmuseum für Vor- und Frühgeschichte, Schleswig, West Germany
Rheinhold & Mahla GmbH, Mannheim, West Germany
Sames Electrostatic, Darmstadt, West Germany
Dr. Slevogt & Co., Weilheim, West Germany

Index

DATE DUE
